STUDY GUIDE FOR VECTOR CALCULUS

THIRD EDITION

BY JERROLD E. MARSDEN AND ANTHONY J. TROMBA

STUDY GUIDE* FOR

VECTOR CALCULUS

THIRD EDITION

BY JERROLD E. MARSDEN AND ANTHONY J. TROMBA

PREPARED BY

KAREN PAO AND FREDERICK SOON

*EVERYTHING YOU WANT TO KNOW TO DO HOMEWORK

W. H. FREEMAN AND COMPANY
NEW YORK

A computer generated image of Enneper's minimal surface. The image was created by James T. Hoffman at the University of Massachusetts, Amherst, using the facilities of the Geometry Analysis Numerics and Graphics Group. Copyright 1986 James T. Hoffman.

ISBN 0-7167-1980-0

Printed in the United States of America

7 8 9 VB 9 9 8 7 6 5 4 3

CONTENTS

HOW TO USE THIS BOOK

This study guide is intended to aid your understanding of vector calculus. We have organized it into chapters and sections corresponding to Marsden and Tromba's *Vector Calculus,* Third Edition. Each section contains Goals, Study Hints, and (most important) Solutions to Selected Exercises. In addition, we have written four sample tests.

The Goals are a short summary of what you ought to learn in the section and what you should understand before going on to the next one. The Goals should also help you review for examinations.

The Study Hints are definitions and facts that you should keep in mind when you do your homework. Here, too, we warn you about mistakes that are commonly made.

For solutions, we have picked some exercises and worked them out. Sometimes we ask you to verify something or to fill in a detail, but most of our solutions are as complete as possible. However, we did not work the problems out so that you can copy them and hand them in as your own work. That's called cheating. We (as well as your instructors) believe that mathematics is not a spectator sport. To understand what's going on, you must do exercises. If you are lost (or have fallen asleep in classes, as one of us has always done), working through our detailed solutions can help you to find your way back. If you feel insecure before exams or quizzes, doing extra exercise problems and comparing your answers with ours is the best way to study. If you are simply studious and want to do extra exercise problems, you will not have to do them blindly, because we have provided many solutions.

Even if you do no more than just flip through the pages of our little book ten minutes before the exam, you should feel better about vector calculus!

We wish you great success!

ACKNOWLEDGMENTS

We would like to thank Professors Marsden and Tromba for giving us the opportunity to write this book. We are especially grateful to Professor Marsden, who let us use his Macintosh Plus and all the necessary software and accessories, meticulously read through our manuscripts, and made many valuable corrections and suggestions. We would also like to thank Andrew Hwang, who provided many solutions, and Sean Bates, for his contributions.

Karen Pao
Frederick Soon

CHAPTER 1
THE GEOMETRY OF EUCLIDEAN SPACE

1.1 Vectors in Three-Dimensional Space

GOALS

1. Be able to perform the following operations with vectors: addition, subtraction, scalar multiplication.

2. Given a vector and a point, be able to write the equation of the line passing through the point in the direction of the vector.

3. Given two points, be able to write the equation of the line passing through them.

STUDY HINTS

1. *Space notation.* The symbol **R** or $\mathbf{R}^1$ refers to all points on the real number line or a one-dimensional space. $\mathbf{R}^2$ refers to all ordered pairs (x, y) which lie in the plane, a two-dimensional space. $\mathbf{R}^3$ refers to all ordered triples (x, y, z) which lie in three-dimensional space. In general, the "exponent" in $\mathbf{R}^n$ tells you how many components there are in each vector.

2. *Vectors and scalars.* A vector has both length (magnitude) and direction. Scalars are just numbers. Scalars do not have direction. Two vectors are equal if and only if they both have the same length and the same direction. Pictorially, they do not need to originate from the same starting point. The vectors shown here are equal.

3. *Vector notation.* Vectors are often denoted by boldface letters, underlined letters, arrows over letters, or by an n-tuple $(x_1, x_2, \ldots, x_n)$. Each x_i of the n-tuple is called the ith component. BEWARE that the n-tuple may represent either a point *or* a vector. The vector (0, 0, ... , 0) is denoted **0** . Your instructor or other textbooks may use other notations such as a squiggly line (~) underneath a letter. A circumflex (^) over a letter is sometimes used to represent a *unit* vector.

4. *Vector addition.* Vectors may be added componentwise, e.g., in $\mathbf{R}^2$
$$(x_1, y_1) + (x_2, y_2) = (x_1 + x_2, y_1 + y_2) .$$
Pictorially, two vectors may be thought of as the sides of a parallelogram. Starting from the vertex formed by the two vectors, we form a new vector which ends at the opposite corner of the

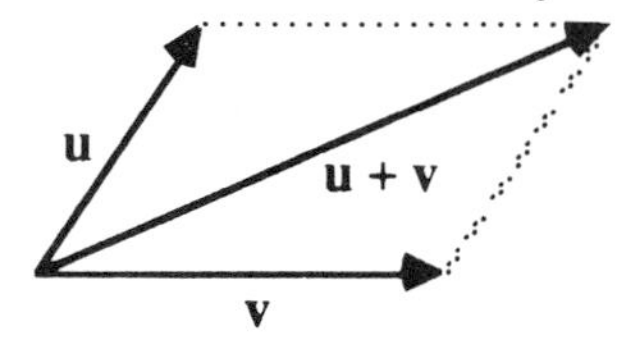

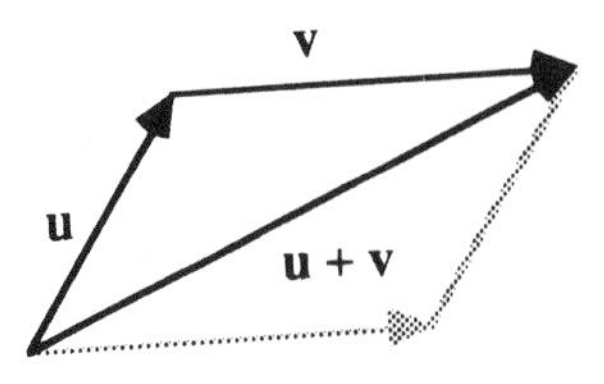

parallelogram. This new vector is the sum of the other two. Alternatively, one could simply translate **v** so that the tail of **v** meets the head of **u**. The vector joining the tail of **u** to the head of **v** is **u** + **v**.

5. *Vector subtraction.* Just as with addition, vectors may be subtracted componentwise.

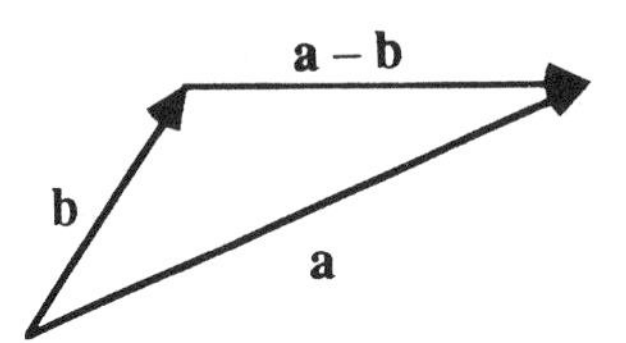

Think of this as adding a negative vector. Pictorially, the vectors **a**, **b**, and **a** – **b** form a triangle. To determine the correct direction, you should be able to add **a** – **b** and **b** to get **a**. Thus, **a** – **b** goes from the tip of **b** to the tip of **a**.

6. *Scalar multiplication.* Here, each component of a vector is multiplied by the same scalar, e.g., in $\mathbf{R}^2$

   $r(x, y) = (rx, ry)$ for any real number r.

   The effect of multiplication by a positive scalar is to change the length by a factor. If the scalar is negative, the lengthening occurs in the opposite direction. Multiplication of vectors will be discussed in the next two sections.

7. *Standard basis vectors.* These are vectors whose components are all 0 except for a single 1. In $\mathbf{R}^3$, **i**, **j**, and **k** denote the vectors which lie on the x, y, and z axes. They are (1, 0, 0), (0, 1, 0), and (0, 0, 1), respectively. The standard basis vectors in $\mathbf{R}^2$ are **i** and **j**, which are vectors lying on the x and y axes, and their respective components are (1, 0) and (0, 1). Sometimes, these vectors are denoted by

   $\mathbf{e}_x, \mathbf{e}_y, \mathbf{e}_z$ or $\mathbf{e}_1, \mathbf{e}_2, \mathbf{e}_3$.

8. *Lines.* (a) The line passing through **a** in the direction of **v** is $\mathbf{l}(t) = \mathbf{a} + t\mathbf{v}$. This is called the point-direction form of the line because the only necessary information is the point **a** and the direction of **v**.

   (b) The line passing through **a** and **b** is $\mathbf{l}(t) = \mathbf{a} + t(\mathbf{b} - \mathbf{a})$. This is called the point-point form of the line. To see if the direction is correct, plug in $t = 0$ and you should get the first point. Plug in $t = 1$ and you should get the second point.

9. *Spanning a space.* If all points in a space can be written in the form $\lambda_1\mathbf{v}_1 + \lambda_2\mathbf{v}_2 + \dots + \lambda_n\mathbf{v}_n$, where λ_i are scalars, then the vectors $\mathbf{v}_1, \dots, \mathbf{v}_n$ span that given space. For example, the vectors **i** and **j** span the xy plane.

10. *Geometric proofs.* The use of vectors can often simplify a proof. Try to compare vector methods and non-vector methods by doing example 7 without vectors.

11. *Problem solving.* Since vectors have magnitude and direction, they can be represented pictorially. Thus, it is often useful to sketch a diagram to help you visualize a vector word problem.

SOLUTIONS TO SELECTED EXERCISES

1. We must solve the following equations:
   $-21 - x = -25$
   $23 - 6 = y$.
   We get $x = 4$ and $y = 17$, so $(-21, 23) - (4, 6) = (-25, 17)$.

4. Convert $-4\mathbf{i} + 3\mathbf{j}$ to $(-4, 3, 0)$, so
   $(2, 3, 5) - 4\mathbf{i} + 3\mathbf{j} = (2, 3, 5) + (-4, 3, 0) = (-2, 6, 5)$.

7. 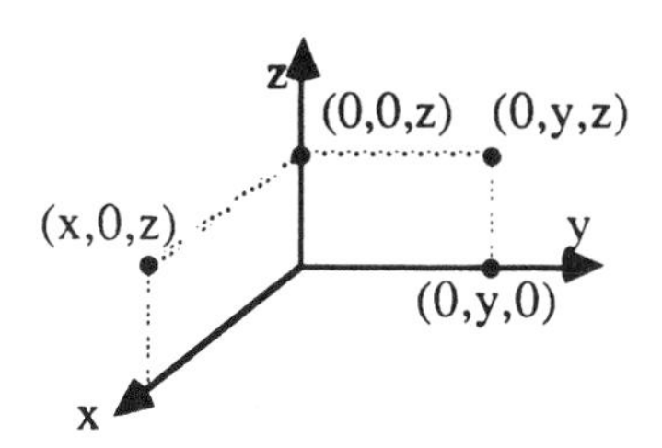

On the y axis, points have the coordinates $(0, y, 0)$, so we must restrict x and z to be 0. On the z axis, points have the coordinates $(0, 0, z)$, so we must restrict x and y to be 0. In the xz plane, points have the coordinates $(x, 0, z)$, so we must restrict y to be 0. In the yz plane, points have the coordinates $(0, y, z)$, so we must restrict x to be 0.

12. Every point on the plane spanned by the given vectors can be written as $a\mathbf{v}_1 + b\mathbf{v}_2$, where a and b are real numbers; therefore, the plane is described by
    $a(3, -1, 1) + b(0, 3, 4)$.

15. Given two points $\mathbf{a}$ and $\mathbf{b}$, a line through them is $\mathbf{l}(t) = \mathbf{a} + t(\mathbf{b} - \mathbf{a})$. In this case, we get
    $\mathbf{l}(t) = (-1, -1, -1) + t(2, 0, 3) = (2t - 1, -1, 3t - 1)$.

19. Substitute $\mathbf{v} = (x, y, z) = (2 + t, -2 + t, -1 + t)$ into the equation for x, y, and z and get
    $2x - 3y + z - 2 =$
    $2(2 + t) - 3(-2 + t) + (-1 + t) - 2 =$
    $4 + 2t + 6 - 3t - 1 + t - 2 = 7$.
    Since $7 \neq 0$, there are no points (x, y, z) satisfying the equation and lying on $\mathbf{v}$.

21. 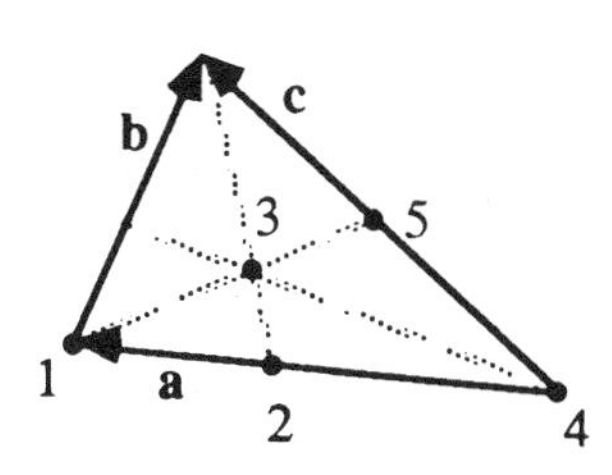

Let $\mathbf{a}$, $\mathbf{b}$, and $\mathbf{c}$ be the sides of the triangle as shown. We assume that each median is divided into a ratio of 2:1 by the point of intersection. The vector from 1 to 2 is $-\mathbf{a}/2 = -(\mathbf{c} - \mathbf{b})/2$. The vector from 2 to 3 is $(1/3)(\mathbf{a}/2 + \mathbf{b})$. The vector from 3 to 4 is $(-2/3)(\mathbf{a} + \mathbf{b}/2)$. The vector from 4 to 5 is $(\mathbf{a} + \mathbf{b})/2$. If our assumption is correct, then the vector from 1 to 5 should be the same as the vector going from 1 to 2 to 3 to 4 to 5 . That is,

$$-\frac{\mathbf{c} - \mathbf{b}}{2} + \frac{1}{3}\left(\frac{\mathbf{a}}{2} + \mathbf{b}\right) - \frac{2}{3}\left(\mathbf{a} + \frac{\mathbf{b}}{2}\right) + \frac{\mathbf{a} + \mathbf{b}}{2} ,$$

which is $\mathbf{b} - \mathbf{c}/2$. This is indeed the vector from 1 to 5 , and it is the median which ends on $\mathbf{c}$. Therefore, our assumption was correct. The other two medians are analyzed in the same way.

22. Just as the parallelogram of example 4 was described by $\mathbf{v} = s\mathbf{a} + t\mathbf{b}$ for s and t in [0, 1], the parallelepiped can be described by $\mathbf{w} = s\mathbf{a} + t\mathbf{b} + r\mathbf{c}$ for s, t, and r in [0, 1].

27. (a) Looking at the x components, the pilot needs to get from 3 to 23. His velocity in the x direction is the **i** component, 400 km/hr. Thus,
$$\Delta t = \frac{\Delta d}{v} = \frac{23-3}{400} = \frac{1}{20}.$$
The pilot flies over the airport (1/20) hour or 3 minutes later. The same answer could have been obtained by analyzing the y components.

(b) We look at the z components and use the formula
$$\Delta d = v\,\Delta t\text{, i.e., } h - 5 = (-1)(1/20).$$
Solving for h, we see that the pilot is (404/20) km above the airport when he passes over.

30. (a)

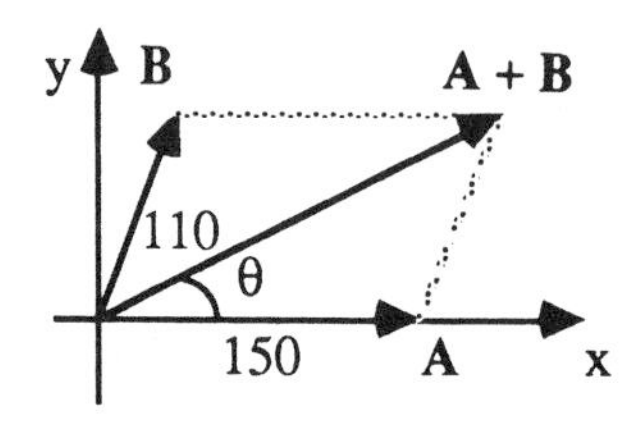

It is convenient to draw the diagram with **A** on the x axis.

(b) From the diagram in part (a), we get
$$\mathbf{A} = 150\mathbf{i} \quad\text{and}\quad \mathbf{B} = (110\cos 60°)\mathbf{i} + (110\sin 60°)\mathbf{j}.$$
$$\mathbf{A} + \mathbf{B} = (150 + 110\cos 60°)\mathbf{i} + (110\sin 60°)\mathbf{j} = 205\mathbf{i} + 55\sqrt{3}\,\mathbf{j}.$$
The angle that $\mathbf{A} + \mathbf{B}$ makes with **A** is
$$\theta = \tan^{-1}\left(\frac{y}{x}\right) = \tan^{-1}\left(\frac{55\sqrt{3}}{205}\right) \approx \tan^{-1}(0.4647) \approx 25°.$$

33. (a) Using the coordinates (C, H, O), we get
$$p(3, 4, 3) + q(0, 0, 2) = r(1, 0, 2) + s(0, 2, 1).$$

(b) To find the smallest integer solution for p, q, r, and s, we balance the equation for the coordinates C, H, and O:

$3p = r$ (equating C)

$2s = 4p$ or $s = 2p$ (equating H)

$2q + 3p = 2r + s = 6p + 2p$ or $q = (5/2)p$. (equating O)

Let p = 2, then the smallest integer solution is
$$p = 2,\ q = 5,\ r = 6,\ s = 4.$$

(c)

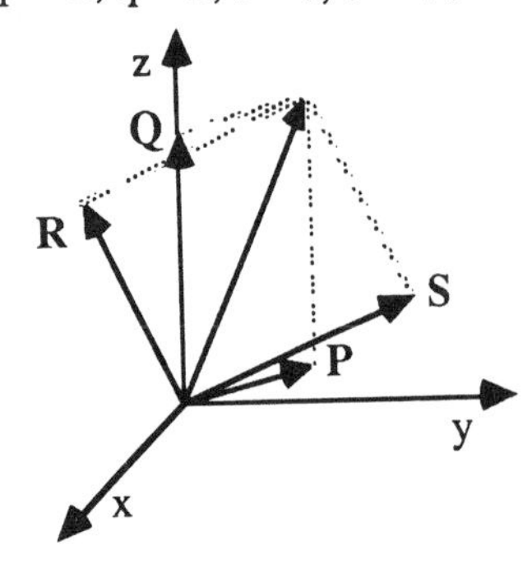

In the diagram, **P** is (6, 8, 6), **Q** is (0, 0, 10), **R** is (6, 0, 12), and **S** is (0, 8, 4). Both sides of the equation add up to the vector (6, 8, 16).

1.2 The Inner Product

GOALS

1. Be able to compute a dot product.

2. Be able to explain the geometric significance of the dot product.

3. Be able to normalize a vector.

4. Be able to compute the projection of one vector onto another.

STUDY HINTS

1. *Inner product.* This is also commonly called the dot product, and it is denoted by $\mathbf{a} \cdot \mathbf{b}$ or $\langle \mathbf{a}, \mathbf{b} \rangle$. The dot product is the sum $\sum_{i=1}^{n} a_i b_i$, where a_i and b_i are the ith components of $\mathbf{a}$ and $\mathbf{b}$, respectively. For example, in $\mathbf{R}^2$, $\mathbf{a} \cdot \mathbf{b} = a_1 b_1 + a_2 b_2$. Note that the dot product is a *scalar.*

2. *Length of a vector.* The length or the norm of a vector $\mathbf{x} = (x, y, z)$ is $\sqrt{x^2 + y^2 + z^2}$. It is denoted by $\| \mathbf{x} \|$ and is equal to $(\mathbf{x} \cdot \mathbf{x})^{1/2}$. This is derivable from the fact that $\mathbf{x} \cdot \mathbf{y} = x_1 y_1 + x_2 y_2 + x_3 y_3$ with $\mathbf{x} = \mathbf{y}$.

3. *Unit vector.* These vectors have length 1. You can make any non-zero vector a unit vector by normalizing it. To normalize a vector, divide the vector by its length, i.e., compute $\mathbf{a} / \| \mathbf{a} \|$.

4. *Cauchy-Schwarz inequality.* Knowing that $| \mathbf{a} \cdot \mathbf{b} | \leq \| \mathbf{a} \| \| \mathbf{b} \|$ is most important for doing proofs in the optional sections of this text and in more advanced courses.

5. *Important geometric properties.* Know that $\mathbf{a} \cdot \mathbf{b} = \| \mathbf{a} \| \| \mathbf{b} \| \cos \theta$, where θ is the angle between the two vectors. As a consequence, $\mathbf{a} \cdot \mathbf{b} = 0$ implies that $\mathbf{a}$ and $\mathbf{b}$ are orthogonal. The zero vector is orthogonal to all vectors.

6. *Projections.* The projection of $\mathbf{b}$ onto $\mathbf{a}$ is the "shadow" of $\mathbf{b}$ falling onto $\mathbf{a}$. The projection of $\mathbf{b}$ onto $\mathbf{a}$ is a vector of length $\mathbf{a} \cdot \mathbf{b}$, in the *direction* of $\mathbf{a}$, or $(\mathbf{a} \cdot \mathbf{b})\mathbf{a} / \| \mathbf{a} \|$.

SOLUTIONS TO SELECTED EXERCISES

3. From the definition of the dot product, we get

$$\cos\theta = \frac{\mathbf{u}\cdot\mathbf{v}}{\|\mathbf{u}\|\,\|\mathbf{v}\|} = \frac{(0, 7, 19)\cdot(-2, -1, 0)}{\sqrt{7^2+19^2}\sqrt{2^2+1^2}} = \frac{-7}{\sqrt{410}\,\sqrt{5}} \approx -0.1546\ .$$

From a hand calculator, we find that $\theta \approx 99°$.

7. If $\mathbf{w} = a\mathbf{i} + b\mathbf{j} + c\mathbf{k}$, then $\|\mathbf{w}\| = [a^2 + b^2 + c^2]^{1/2}$, and so

$$\|\mathbf{u}\| = \sqrt{1+4} = \sqrt{5}\ ; \quad \|\mathbf{v}\| = \sqrt{1+1} = \sqrt{2}\ .$$

Using the formula for the dot product, we get

$$\mathbf{u}\cdot\mathbf{v} = (-1)(1) + (2)(-1) = -3\ .$$

10. Using the same formulas as in exercise 7, we get

$$\|\mathbf{u}\| = \sqrt{1+0+9} = \sqrt{10}\ ; \quad \|\mathbf{v}\| = \sqrt{0+16+0} = 4\ .$$

Since $\mathbf{u}$ does not have a $\mathbf{j}$ component and $\mathbf{v}$ does not have any $\mathbf{i}$ or $\mathbf{k}$ component, the vectors are perpendicular; therefore, $\mathbf{u}\cdot\mathbf{v} = 0$.

12. A vector $\mathbf{w}$ is normalized by the formula $\mathbf{w}/\|\mathbf{w}\|$. For the vectors in exercise 7:

$$\frac{\mathbf{u}}{\|\mathbf{u}\|} = \frac{1}{\sqrt{5}}(-\mathbf{i} + 2\mathbf{j})\ ; \quad \frac{\mathbf{v}}{\|\mathbf{v}\|} = \frac{1}{\sqrt{2}}(\mathbf{i} - \mathbf{j})\ .$$

15. The projection of $\mathbf{v}$ onto $\mathbf{u}$ is

$$\frac{\mathbf{u}\cdot\mathbf{v}}{\|\mathbf{u}\|^2}\,\mathbf{u} = \frac{(-1)(2) + (1)(1) + (1)(-3)}{1+1+1}\,(-\mathbf{i}+\mathbf{j}+\mathbf{k}) = \frac{-4}{3}(-\mathbf{i}+\mathbf{j}+\mathbf{k})\ .$$

16. For orthogonality, we want the dot product to be 0 .
(a) The dot product is $(2\mathbf{i} + b\mathbf{j})\cdot(-3\mathbf{i} + 2\mathbf{j} + \mathbf{k}) = -6 + 2b$, so b must be 3 .
(b) The dot product is $(2\mathbf{i} + b\mathbf{j})\cdot\mathbf{k} = 0$, so b can be any real number.

21. (a) Geometrically, we see that the $\mathbf{i}$ component of $\mathbf{F}$ is $\|\mathbf{F}\|\cos\theta$. Similarly, the $\mathbf{j}$ component of $\mathbf{F}$ is $\|\mathbf{F}\|\sin\theta$. Therefore, $\mathbf{F} = \|\mathbf{F}\|\cos\theta\,\mathbf{i} + \|\mathbf{F}\|\sin\theta\,\mathbf{j}$, where θ is the angle from the x axis. Since the angle from the y axis is $\pi/4$, θ is also $\pi/4$, so $\mathbf{F} = 3\sqrt{2}\,(\mathbf{i} + \mathbf{j})$.

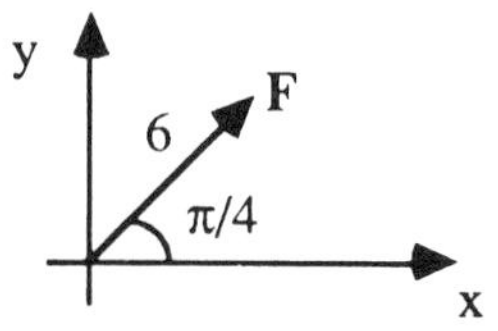

(b) We compute $\mathbf{D} = 4\mathbf{i} + 2\mathbf{j}$, so $\mathbf{F}\cdot\mathbf{D} = (3\sqrt{2})(4) + (3\sqrt{2})(2) = 18\sqrt{2}$. Also, $\|\mathbf{F}\| = 6$ and $\|\mathbf{D}\| = \sqrt{20}$. From the definition of the dot product,

$$\cos\theta = \frac{\mathbf{F}\cdot\mathbf{D}}{\|\mathbf{F}\|\,\|\mathbf{D}\|} = \frac{18\sqrt{2}}{6\sqrt{20}} = \frac{3}{\sqrt{10}} \approx 0.9487\ .$$

Thus, $\theta \approx 18°$.

(c) From part (b), we had computed $\mathbf{F}\cdot\mathbf{D} = 18\sqrt{2}$. Knowing that $\cos\theta = 3/\sqrt{10}$, we calculate $\|\mathbf{F}\|\,\|\mathbf{D}\|\cos\theta = (6)\sqrt{20}\,(3/\sqrt{10}) = 18\sqrt{2}$, also.

1.3 The Cross Product

GOALS

1. Be able to compute a cross product.

2. Be able to explain the geometric significance of the cross product.

3. Be able to write the equation of a plane from given information regarding points on the plane or normals to the plane.

STUDY HINTS

1. *Matrices and determinants.* A matrix is just a rectangular array of numbers. The array is written between a set of brackets. The determinant of a matrix is a number; a matrix has no numerical value. The determinant is defined only for square matrices and it is denoted by vertical bars.

2. *Computing determinants.* Know that $\begin{vmatrix} a & b \\ c & d \end{vmatrix} = ad - bc$. Also know that

$$\begin{vmatrix} a & b & c \\ d & e & f \\ g & h & i \end{vmatrix} = a\begin{vmatrix} e & f \\ h & i \end{vmatrix} - b\begin{vmatrix} d & f \\ g & i \end{vmatrix} + c\begin{vmatrix} d & e \\ g & h \end{vmatrix}.$$

Note the minus sign in front of the second term on the right-hand side. The general method for computing determinants is described below in item 3.

3. *Computing* $n \times n$ *determinants.* Use the checkerboard pattern shown here which begins with a plus sign in the upper left corner. Choose any column or row -- usually picking the one with the most zeroes saves work. Draw vertical and horizontal lines through the first number of the row or column. The numbers remaining form an $n-1 \times n-1$ determinant, which should be multiplied by the number (with sign determined by the checkerboard) through which both lines are drawn. Repeat for the remaining numbers of the row or column. Finally, sum the results. This process, called expansion by minors, works for any row or column. The best way to remember the process is by practicing. Be sure to use the correct signs.

$$\begin{vmatrix} + & - & + & \cdots \\ - & + & - & \cdots \\ + & - & + & \cdots \\ \vdots & & & \vdots \end{vmatrix}$$

4. *Simplifying determinants.* Determinants are easiest to compute when zeroes are present. Adding one row or one column to another row or column does not change the value of a determinant, but this can simplify the computation. See example 3.

5. *Computing a cross product.* If $\mathbf{a} = (a_1, a_2, a_3)$ and $\mathbf{b} = (b_1, b_2, b_3)$, then

$$\mathbf{a} \times \mathbf{b} = \begin{vmatrix} \mathbf{i} & \mathbf{j} & \mathbf{k} \\ a_1 & a_2 & a_3 \\ b_1 & b_2 & b_3 \end{vmatrix}$$

The order matters: $\mathbf{a} \times \mathbf{b} = -(\mathbf{b} \times \mathbf{a})$. The cross product is *not* commutative. Also, note that $\mathbf{a} \times \mathbf{b}$ is a vector, *not* a scalar.

6. *Properties of the cross product.* The vectors $\mathbf{a}$, $\mathbf{b}$, and $\mathbf{a} \times \mathbf{b}$ form a right-handed system (see figure 1.3.2). The cross product $\mathbf{a} \times \mathbf{b}$ is orthogonal to both $\mathbf{a}$ and $\mathbf{b}$. The length of $\mathbf{a} \times \mathbf{b}$ is $\|\mathbf{a}\| \|\mathbf{b}\| |\sin\theta|$, where θ is the angle between $\mathbf{a}$ and $\mathbf{b}$. Note that the cross product is related to $\sin\theta$, whereas the dot product is related to $\cos\theta$.

7. *More properties.* If the cross product is zero, then either: (i) the length of one of the vectors must be zero, or (ii) $\sin\theta = 0$, i.e., $\theta = 0$, so the vectors must be parallel.

8. *Geometry.* The absolute value of the determinant $\begin{vmatrix} a & b \\ c & d \end{vmatrix}$ is the area of the parallelogram spanned by the vectors (a, b) and (c, d) originating from the same point. The absolute value of the determinant $\begin{vmatrix} a & b & c \\ d & e & f \\ g & h & i \end{vmatrix}$ is the volume of the parallelepiped spanned by the vectors (a, b, c), (d, e, f), and (g, h, i) originating from the same point. The length of the cross product $\|\mathbf{a} \times \mathbf{b}\|$ is the area of the parallelogram spanned by the vectors $\mathbf{a}$ and $\mathbf{b}$. The vector $\mathbf{a} \times \mathbf{b}$ gives a vector normal to the plane spanned by $\mathbf{a}$ and $\mathbf{b}$.

9. *Plane equation.* Recall that the equation of a plane is $ax + by + cz + d = 0$. The vector (a, b, c) is orthogonal to the plane. Knowing two vectors in the plane, we can determine an orthogonal vector by using the cross product. Compare methods 1 and 2 of example 8.

10. *Distance from point to plane.* You should understand the derivation of the equation in example 9. If necessary, review the geometric properties of the dot product in section 1.2.

SOLUTIONS TO SELECTED EXERCISES

2. (b) We subtract 12 times the third row from the first row and subtract 15 times the third row from the second row. Then we expand along the first column:

$$\begin{vmatrix} 36 & 18 & 17 \\ 45 & 24 & 20 \\ 3 & 5 & -2 \end{vmatrix} = \begin{vmatrix} 0 & -42 & 41 \\ 0 & -51 & 50 \\ 3 & 5 & -2 \end{vmatrix} = 3\begin{vmatrix} -42 & 41 \\ -51 & 50 \end{vmatrix} = 3(-2100 + 2091) = -27 .$$

5. The area of the parallelogram is $\|\mathbf{a}\times\mathbf{b}\|$. We compute

$$\mathbf{a}\times\mathbf{b} = \begin{vmatrix} \mathbf{i} & \mathbf{j} & \mathbf{k} \\ 1 & -2 & 1 \\ 2 & 1 & 1 \end{vmatrix} = -3\mathbf{i} + \mathbf{j} + 5\mathbf{k}$$

and so the area of the parallelogram is

$$\|\mathbf{a}\times\mathbf{b}\| = \sqrt{9+1+25} = \sqrt{35}\ .$$

8. The volume of the parallelepiped is the absolute value of the 3×3 determinant made up of the vectors' components. Expand along the first row:

$$\begin{vmatrix} 1 & 0 & 0 \\ 0 & 3 & -1 \\ 4 & 2 & -1 \end{vmatrix} = \begin{vmatrix} 3 & -1 \\ 2 & -1 \end{vmatrix} = -1\ .$$

Thus, the volume is 1.

11. We want to find the cross product and then normalize it. Since $\mathbf{0}$ is orthogonal to any vector, we may ignore it. We compute:

$$\mathbf{v} = \begin{vmatrix} \mathbf{i} & \mathbf{j} & \mathbf{k} \\ -5 & 9 & -4 \\ 7 & 8 & 9 \end{vmatrix} = 113\mathbf{i} + 17\mathbf{j} - 103\mathbf{k}\ .$$

So

$$\|\mathbf{v}\| = \sqrt{113^2 + 17^2 + 103^2} = \sqrt{23667}\ .$$

There are two orthogonal vectors in opposing directions, they are $\pm\,\mathbf{v}/\|\mathbf{v}\| =$ $\pm(113\mathbf{i} + 17\mathbf{j} - 103\mathbf{k})/\sqrt{23667}$.

15. (a) The equation of a plane with normal vector (A, B, C) and passing through the point (x_0, y_0, z_0) is $A(x - x_0) + B(y - y_0) + C(z - z_0) = 0$. In this case, the equation is

$$1(x-1) + 1(y-0) + 1(z-0) = 0 \quad \text{or} \quad x + y + z = 1\ .$$

(d) Here, the normal vector is parallel to the line, so it is $(-1, -2, 3)$. Hence, the equation of the desired plane is

$$-1(x-2) - 2(y-4) + 3(z+1) = 0 \quad \text{or} \quad -x - 2y + 3z + 13 = 0\ .$$

16. (b) Two vectors in the desired plane are $\mathbf{v} = (0-1, 1-2, -2-0) = (-1, -1, -2)$ and $\mathbf{w} = (4-0, 0-1, 1+2) = (4, -1, 3)$. The cross product $\mathbf{v}\times\mathbf{w}$ is orthogonal to both vectors and hence, normal to the desired plane. We compute:

$$\mathbf{v}\times\mathbf{w} = -5\mathbf{i} - 5\mathbf{j} + 5\mathbf{k}\ ,$$

so the desired equation is

$$-5(x-1) - 5(y-2) + 5(z-0) = 0 \quad \text{or} \quad -x - y + z + 3 = 0\ .$$

18. (a) Let D be the matrix with rows $\mathbf{u}, \mathbf{v}, \mathbf{w}$, then

$$\det D = \mathbf{u} \cdot (\mathbf{v} \times \mathbf{w}) = \begin{vmatrix} u_1 & u_2 & u_3 \\ v_1 & v_2 & v_3 \\ w_1 & w_2 & w_3 \end{vmatrix} .$$

Use the following property of determinants: $\mathbf{v} \cdot (\mathbf{w} \times \mathbf{u})$ corresponds to two row exchanges of the matrix D, so we have

$$\mathbf{v} \cdot (\mathbf{w} \times \mathbf{u}) = (-1)(-1) \det D = \det D \quad \text{and}$$

$$\mathbf{w} \cdot (\mathbf{u} \times \mathbf{v}) = (-1)(-1) \det D = \det D .$$

To prove the other three, recall that $\mathbf{u} \times \mathbf{v} = -(\mathbf{v} \times \mathbf{u})$.

22. The line perpendicular to the plane is the line parallel to the normal of the plane, so the equation of the line is

$$\mathbf{l}(t) = (1, -2, -3) + t(3, -1, -2) .$$

25. Let $\mathbf{u}$ be the vector normal to the plane. Then $\mathbf{u}$ is perpendicular to $3\mathbf{i} + 2\mathbf{j} + 4\mathbf{k}$ since $\mathbf{v}$ is on the plane. Also, $\mathbf{u}$ is perpendicualr to $2\mathbf{i} + \mathbf{j} - 3\mathbf{k}$ because the vector $A\mathbf{i} + B\mathbf{j} + C\mathbf{k}$ is perpendicular all vectors in the plane $Ax + By + Cz = D$. To find $\mathbf{u}$, we take the cross product of $3\mathbf{i} + 2\mathbf{j} + 4\mathbf{k}$ and $2\mathbf{i} + \mathbf{j} - 3\mathbf{k}$:

$$\mathbf{u} = \begin{vmatrix} \mathbf{i} & \mathbf{j} & \mathbf{k} \\ 3 & 2 & 4 \\ 2 & 1 & -3 \end{vmatrix} = -10\mathbf{i} + 17\mathbf{j} - \mathbf{k} .$$

When $t = 0$, we find that a point on the plane is $(-1, 1, 2)$, so the equation of the plane is

$$-10(x + 1) + 17(y - 1) - (z - 2) = 0 \quad \text{or} \quad -10x + 17y - z + 25 = 0 .$$

26. First, find the normal of the plane. The normal of the plane is perpendicular to the *line* passing through $(3, 2, -1)$ and $(1, -1, 2)$. The equation of the line is $\mathbf{l}(t) = (3, 2, -1) + t(2, 3, -3)$. The normal of the plane is also perpendicular to $\mathbf{v} = (1, -1, 0) + t(3, 2, -2)$. Therefore, the normal is $(2\mathbf{i} + 3\mathbf{j} - 3\mathbf{k}) \times (3\mathbf{i} + 2\mathbf{j} - 2\mathbf{k}) = -5\mathbf{j} + 5\mathbf{k}$. Now, we need a point on the plane, say, $(3, 2, -1)$. Thus, the equation of the plane is

$$0(x - 3) - 5(y - 2) + 5(z + 1) = 0 \quad \text{or} \quad -y + z = 1 .$$

30. The plane passing through the origin and perpendicular to $\mathbf{i} - 2\mathbf{j} + \mathbf{k}$ is $x - 2y + z = 0$. By the distance formula with $(A, B, C, D) = (1, -2, 1, 0)$ and $(x_1, y_1, z_1) = (6, 1, 0)$,

$$d = \frac{|Ax_1 + By_1 + Cz_1 + D|}{\sqrt{A^2 + B^2 + C^2}} = \frac{6 - 2}{\sqrt{1 + 4 + 1}} = \frac{4}{\sqrt{6}} = \frac{2\sqrt{6}}{3} .$$

33. Since all vectors in this exercise are unit vectors, $\|\mathbf{N} \times \mathbf{a}\| = \sin\theta_1$ and $\|\mathbf{N} \times \mathbf{b}\| = \sin\theta_2$. From Snell's law, $n_1 \sin\theta_1 = n_2 \sin\theta_2$. Hence,

$$n_1 \|\mathbf{N} \times \mathbf{a}\| = n_2 \|\mathbf{N} \times \mathbf{b}\| .$$

In order to establish that $\mathbf{N} \times \mathbf{a}$ and $\mathbf{N} \times \mathbf{b}$ have the same direction, we assume that $\mathbf{N}$, $\mathbf{a}$, and $\mathbf{b}$ all lie in the same plane, and $\mathbf{a}$ and $\mathbf{b}$ are on the same side of $\mathbf{N}$.

Hence $\mathbf{N} \times \mathbf{a}$ and $\mathbf{N} \times \mathbf{b}$ both are perpendicular to this plane and parallel to each other. Thus, $n_1 \; \|\mathbf{N} \times \mathbf{a}\|$ and $n_2 \; \|\mathbf{N} \times \mathbf{b}\|$ are equal.

34. First, 4 times the first row is subtracted from the second row. Next, 7 times the first row is subtracted from the third row. The next step is expansion by minors along the first column. Finally, the 2×2 determinant is computed.

1.4 Cylindrical and Spherical Coordinates

GOALS

1. Be able to convert back and forth between the cylindrical, spherical, and cartesian coordinate systems.

2. Be able to describe geometric objects with cylindrical and spherical coordinates.

3. Be able to describe the geometric effects of changing a coordinate.

STUDY HINTS

1. *Review.* You should review polar coordinates in your one-variable calculus text.

2. *Cylindrical coordinates.* Denoted (r, θ, z), this is just like polar coordinates except that a z coordinate has been added. Know the formulas $x = r \cos \theta$, $y = r \sin \theta$, $r = \sqrt{x^2 + y^2}$, and $\theta = \tan^{-1}(y/x)$.

3. *Spherical coordinates.* Denoted by (ρ, θ, φ) , ρ is the distance from the origin, φ is the angle from the positive z axis, and θ is the same as in cylindrical coordinates. Know the formulas $x = \rho \cos \theta \sin \varphi$, $y = \rho \sin \theta \sin \varphi$, and $z = \rho \cos \varphi$. Also know that $\rho = \sqrt{x^2 + y^2 + z^2}$, $\theta = \tan^{-1}(y/x)$, and $\varphi = \cos^{-1}(z/\sqrt{x^2 + y^2 + z^2}) = \cos^{-1}(z/\rho)$.

4. *Graphs of* $r, \rho =$ *constant.* Note that $r =$ constant in cylindrical coordinates describes a cylinder and that $\rho =$ constant in spherical coordinates describes a sphere. You may have suspected this from the name of the coordinate system.

5. *Computing* θ *and* φ . Remember that, in this text, φ takes values from 0 to π and θ ranges from 0 to 2π . In some instances, it is more convenient to define θ in the

range $-\pi$ to π. You should be very careful about computing θ. If $x = y = -1$, then $\tan^{-1}(y/x) = \pi/4$, but plotting $(-1, -1)$ in the xy plane shows that in reality, $\theta = 5\pi/4$. This is why the authors fuss with $\tan^{-1}(y/x)$ in the definition. Plotting the x and y coordinates is very helpful for determining θ.

6. *Negative* r, ρ. Note that we have defined r and ρ to be non-negative. If the distance is given as a negative number, we need to reflect the given point across the origin.

7. *Units vectors in spherical and cylindrical coordinates.* The unit vectors in cylindrical coordinates are $\mathbf{e}_r$, $\mathbf{e}_\theta$, and $\mathbf{e}_z$. $\mathbf{e}_r$ points along the direction of r. $\mathbf{e}_\theta$ goes in the direction in which θ is measured, and $\mathbf{e}_z = \mathbf{k}$. As one might expect, those three unit vectors form an orthonormal basis, and $\mathbf{e}_r \times \mathbf{e}_\theta = \mathbf{e}_z$. Those vectors, however, are not *fixed* as is the case with $\mathbf{i}$, $\mathbf{j}$, and $\mathbf{k}$, that is, if you change the point (r, θ, z), the set of unit vectors *rotate*. For spherical coordinates, there is also a set of unit vectors $\mathbf{e}_\rho$, $\mathbf{e}_\theta$, and $\mathbf{e}_\varphi$. Those vectors, in terms of $\mathbf{i}$, $\mathbf{j}$, $\mathbf{k}$, and the cartesian coordinates of the point are worked out in exercise 7 (see below) of this section.

SOLUTIONS TO SELECTED EXERCISES

1. (a) To convert to rectangular coordinates, use $x = r\cos\theta$ and $y = r\sin\theta$:

$$x = 1\cos 45° = \sqrt{2}/2 \quad \text{and} \quad y = 1\sin 45° = \sqrt{2}/2.$$

Next, use $\rho = \sqrt{x^2 + y^2 + z^2}$ and $\varphi = \cos^{-1}(z/\rho)$ to get the spherical coordinates:

$$\rho = \sqrt{\frac{1}{2} + \frac{1}{2} + 1} = \sqrt{2} \quad \text{and} \quad \varphi = \cos^{-1}\left(\frac{1}{\sqrt{2}}\right) = \frac{\pi}{4}.$$

Hence, for the cylindrical coordinates $(1, 45°, 1)$, the rectangluar coordinates are $(\sqrt{2}/2, \sqrt{2}/2, 1)$ and the spherical coordinates are $(\sqrt{2}, \pi/4, \pi/4)$.

(b) To convert to cylindrical coordinates, we use $r = \sqrt{x^2 + y^2}$ and $\theta = \tan^{-1}(y/x)$:

$$r = \sqrt{4 + 1} = \sqrt{5} \quad \text{and} \quad \theta = \tan^{-1}(1/2).$$

Next, use the same formulas as in part (a) to get the spherical coordinates:

$$\rho = \sqrt{4 + 4 + 1} = 3 \quad \text{and} \quad \varphi = \cos^{-1}(-2/3).$$

Hence, the rectangular coordinates $(2, 1, -2)$ converts to the cylindrical cordinates $(\sqrt{5}, \tan^{-1}(1/2), -2)$ and to the spherical coordinates $(3, \tan^{-1}(1/2), \cos^{-1}(-2/3))$.

2. (b) This mapping takes a point and rotates it by π radians about the z axis. This is followed by a reflection across the xy plane. The net effect is that the point is reflected through the origin.

3. (b) This mapping takes a point and reflects it across the xy plane.

5. Let $\rho \geq 0$, then $(\rho, 0, 0)$ is the positive z axis. Now, let φ vary from 0 to π. Then $(\rho, 0, \varphi)$ is the half-plane in the xz plane with $x \geq 0$. By allowing θ to vary from 0 to 2π, we rotate the half plane described above. Therefore, $\rho \geq 0, 0 \leq \theta < 2\pi$, and $0 \leq \varphi \leq \pi$ describes all points in $\mathbf{R}^3$.
If $\rho < 0$ also, the coordinates are not unique. For example, $(x, y, z) = (1, 0, 0)$ has spherical coordinates $(\sqrt{2}, \pi/4, \pi/2)$ or $(-\sqrt{2}, 5\pi/4, \pi/2)$.

7. (a) First, $\mathbf{e}_\rho$ is the unit vector along the vector (x, y, z); therefore, the formula is

$$\mathbf{e}_\rho = \frac{x\mathbf{i} + y\mathbf{j} + z\mathbf{k}}{\sqrt{x^2 + y^2 + z^2}}.$$

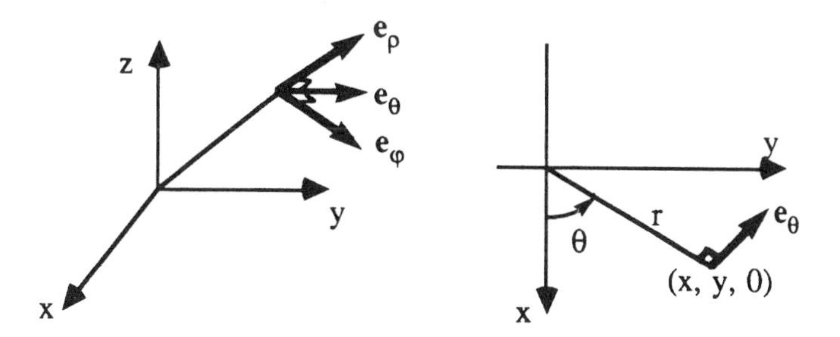

Next, $\mathbf{e}_\theta$ is parallel to the xy plane and denotes the direction in which the angle θ is measured. It is perpendicular to $r = (x, y, 0)$, so $(a\mathbf{i} + b\mathbf{j}) \cdot (x\mathbf{i} + y\mathbf{j}) = 0$. Since we want θ measured counterclockwise, $a = -y$ (instead of y) and $b = x$. Therefore,

$$\mathbf{e}_\theta = \frac{-y\mathbf{i} + x\mathbf{j}}{\sqrt{x^2 + y^2}}.$$

To find $\mathbf{e}_\varphi$, we note that since $\mathbf{e}_\rho$, $\mathbf{e}_\theta$, and $\mathbf{e}_\varphi$ is a set of orthonormal vectors; they form a right-handed coordinate system with $\mathbf{e}_\rho \times \mathbf{e}_\theta = \mathbf{e}_\varphi$. So

$$\mathbf{e}_\varphi = \frac{(x, y, z) \times (-y, x, 0)}{r\rho} = \frac{-xz\mathbf{i} - yz\mathbf{j} + (x^2 + y^2)\mathbf{k}}{\sqrt{x^2 + y^2}\sqrt{x^2 + y^2 + z^2}}.$$

9. (a) The length of $x\mathbf{i} + y\mathbf{j} + z\mathbf{k}$ is $\sqrt{x^2 + y^2 + z^2}$, which is the definition of ρ.
   (b) Note that $\|\mathbf{v}\| = \sqrt{x^2 + y^2 + z^2} = \rho$ and $\mathbf{v} \cdot \mathbf{k} = z$; therefore, $\varphi = \cos^{-1}(z/\rho) = \cos^{-1}(\mathbf{v} \cdot \mathbf{k} / \|\mathbf{v}\|)$.
   (c) Note that $\|\mathbf{u}\| = \sqrt{x^2 + y^2}$, which is the cylindrical coordinate r and $\mathbf{u} \cdot \mathbf{i} = x$; therefore, $\theta = \cos^{-1}(x/r) = \cos^{-1}(\mathbf{u} \cdot \mathbf{i} / \|\mathbf{u}\|)$.

13. Note that φ will be between $\pi/2$ and π because the region lies in the lower

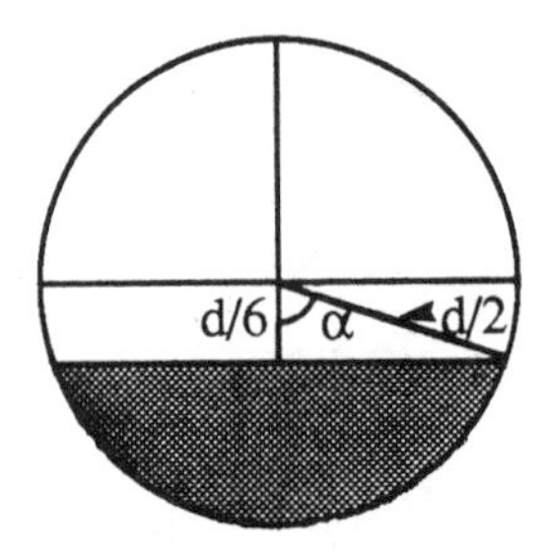

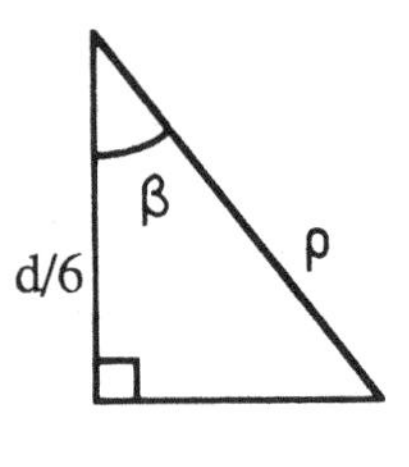

hemisphere. From the triangle, we see that $\cos\alpha = (d/6) \div (d/2) = 1/3$; therefore, we have $\pi - \alpha \leq \varphi \leq \pi$ or $\pi - \cos^{-1}(1/3) \leq \varphi \leq \pi$. Now, ρ can be as large as $d/2$; however, as ρ gets smaller, its lower limit depends on φ . Pick any φ , then $\varphi + \beta = \pi$ and according to the diagram $\cos\beta = (d/6) \div \rho$. Rearrangement gives $d/(6\cos\beta) = \rho = d/(6\cos(\pi - \varphi)) = -d/(6\cos\varphi)$. Therefore, $-d/(6\cos\varphi) \leq \rho \leq d/2$. So far, we have described the cross-section in one quadrant. The entire volume requires a revolution around the z axis, so its description is

$$-\frac{d}{6\cos\varphi} \leq \rho \leq \frac{d}{2}, \quad 0 \leq \theta \leq 2\pi, \text{ and } \pi - \cos^{-1}\left(\frac{1}{3}\right) \leq \varphi \leq \pi .$$

1.5 n-Dimensional Euclidean Space

GOALS

1. Be able to extend the ideas of the previous sections to $\mathbf{R}^n$.

2. Be able to multiply matrices.

STUDY HINTS

1. *The space* $\mathbf{R}^n$. Most of this textbook deals with the Euclidean spaces that we can visualize, $\mathbf{R}^2$ and $\mathbf{R}^3$. Many of the same properties hold in $\mathbf{R}^n$. Vector addtion, scalar multiplication, vector lengths, the dot product, and the triangle inequality are defined similarly.

2. *No cross product analog*. The cross product in $\mathbf{R}^3$ does *not* have an easy analog in $\mathbf{R}^n$, $n \geq 4$.

3. *Standard basis vectors.* The analogs of **i**, **j**, **k** are denoted $\mathbf{e}_i$. The vector $\mathbf{e}_i$ is $(0, 0, \ldots, 1, \ldots, 0)$ with 1 in the ith position. $\mathbf{e}_i$ and $\mathbf{e}_j$ are orthogonal if $i \neq j$.

4. *Matrices.* A matrix is a rectangular array of numbers. Unlike a determinant, a matrix has no numerical value. You should remember that we talk about rows before columns. Thus, an $n \times m$ matrix has n rows and m columns. The (i, j) entry is the number located in row i, column j.

5. *Matrix multiplication.* You should practice until matrix multiplication becomes second nature to you. Let the components of A be a_{ij} and let those of B be b_{kl}, where A is an $m \times p$ matrix and B is a $p \times n$ matrix. Then the components of AB are

$$(ab)_{mn} = \sum_{j=1}^{p} a_{mj}\, b_{jn}\,.$$

We can only multiply an $m \times p$ matrix with a $p \times n$ matrix, i.e., $[m \times p][p \times n]$. Note that the number of columns of A and the number of rows of B must be equal (p in this case).

6. *Non-commutativity of matrix multiplication.* In general, $AB \neq BA$. In fact, AB may be defined when BA is undefined. However, matrix multiplication *is* associative, i.e., $(AB)C = A(BC)$ if the product ABC is defined.

7. *Matrices and mappings.* An $m \times n$ matrix can represent a mapping from $\mathbf{R}^n$ to $\mathbf{R}^m$. To see this, let A be the matrix and let **x** be a vector in $\mathbf{R}^n$, represented as an $n \times 1$ matrix, and **y** is a vector in $\mathbf{R}^m$, an $m \times 1$ matrix. Then the matrix A takes a point in $\mathbf{R}^n$ to a point in $\mathbf{R}^m$ by the equation $A\mathbf{x} = \mathbf{y}$.

SOLUTIONS TO SELECTED EXERCISES

2. (a) Use the properties of lengths and dot products:

$$\begin{aligned}\|\mathbf{x}+\mathbf{y}\|^2 + \|\mathbf{x}-\mathbf{y}\|^2 &= (\mathbf{x}+\mathbf{y})\cdot(\mathbf{x}+\mathbf{y}) + (\mathbf{x}-\mathbf{y})\cdot(\mathbf{x}-\mathbf{y}) \\ &= \mathbf{x}\cdot\mathbf{x} + 2\mathbf{x}\cdot\mathbf{y} + \mathbf{y}\cdot\mathbf{y} + \mathbf{x}\cdot\mathbf{x} - 2\mathbf{x}\cdot\mathbf{y} + \mathbf{y}\cdot\mathbf{y} \\ &= 2\mathbf{x}\cdot\mathbf{x} + 2\mathbf{y}\cdot\mathbf{y} = 2\|\mathbf{x}\|^2 + 2\|\mathbf{y}\|^2\,.\end{aligned}$$

4. To verify the CBS inequality, we compute

$$|\mathbf{x}\cdot\mathbf{y}| = |(1)(3) + (0)(8) + (2)(4) + (6)(1)| = 17\,.$$
$$\|\mathbf{x}\| = \sqrt{1+0+4+36} = \sqrt{41}\,.$$
$$\|\mathbf{y}\| = \sqrt{9+64+16+1} = \sqrt{90}\,.$$

Thus, we indeed have $|\mathbf{x}\cdot\mathbf{y}| = \sqrt{17}\sqrt{17} < \sqrt{41}\sqrt{90} = \|\mathbf{x}\|\,\|\mathbf{y}\|$.
For the triangle inequality, we compute

$\mathbf{x}+\mathbf{y} = (4, 8, 6, 7)$ and

$$\|\mathbf{x}+\mathbf{y}\| = \sqrt{16+64+36+49} = \sqrt{165} < 13\,.$$

Indeed, we have $\|\mathbf{x}+\mathbf{y}\| < 13 < 15 = 6+9 < \sqrt{41} + \sqrt{90} = \|\mathbf{x}\| + \|\mathbf{y}\|$.

8. We compute:

$$AB = \begin{bmatrix} 3 & 1 & -3 \\ 5 & -1 & 1 \\ 1 & 1 & -1 \end{bmatrix} \quad \text{and} \quad A + B = \begin{bmatrix} 4 & 0 & 0 \\ 3 & 2 & 0 \\ 1 & 1 & 1 \end{bmatrix}.$$

Expanding by minors across the first row gives

$$\det A = 3\begin{vmatrix} 2 & -1 \\ 0 & 1 \end{vmatrix} + 1\begin{vmatrix} 1 & 2 \\ 1 & 0 \end{vmatrix} = 6 - 2 = 4,$$

$$\det B = 1\begin{vmatrix} 0 & 1 \\ 1 & 0 \end{vmatrix} - 1\begin{vmatrix} 2 & 0 \\ 0 & 1 \end{vmatrix} = -3,$$

$$\det(AB) = 3\begin{vmatrix} -1 & 1 \\ 1 & -1 \end{vmatrix} - 1\begin{vmatrix} 5 & 1 \\ 1 & -1 \end{vmatrix} - 3\begin{vmatrix} 5 & -1 \\ 1 & 1 \end{vmatrix} = -12, \text{ and}$$

$$\det(A + B) = 4\begin{vmatrix} 2 & 0 \\ 1 & 1 \end{vmatrix} = 8.$$

11. (a) For $n = 2$, $\det(\lambda A) = \begin{vmatrix} \lambda a_{11} & \lambda a_{12} \\ \lambda a_{21} & \lambda a_{22} \end{vmatrix} = \lambda^2(a_{11}a_{22} - a_{12}a_{21}) = \lambda^2 \det A$.

For $n = 3$,
$$\det(\lambda A) = \begin{vmatrix} \lambda a_{11} & \lambda a_{12} & \lambda a_{13} \\ \lambda a_{21} & \lambda a_{22} & \lambda a_{23} \\ \lambda a_{31} & \lambda a_{32} & \lambda a_{33} \end{vmatrix}$$
$$= \lambda a_{11}(\det(\lambda A_1)) - \lambda a_{12}(\det(\lambda A_2)) + \lambda a_{13}(\det(\lambda A_3)),$$
$$= \lambda \cdot \lambda^2 (a_{11} \det A_1 - a_{12} \det A_2 + a_{13} \det A_3)$$
$$= \lambda^3 \det A,$$

where A_1, A_2, and A_3 are 2×2 matrices obtained by expanding across the first row.

Assume that for $n = k$, $\det(\lambda A) = \lambda^k \det A$. Then for $n = k + 1$, $\det(\lambda A)$ can be found by a process analogous to the 3×3 case:

$$\det(\lambda A) = \lambda a_{11}(\det(\lambda A_1)) - \lambda a_{12}(\det(\lambda A_2)) + \dots +$$
$$(-1)^k \lambda a_{1k}(\det(\lambda A_{k+1})),$$
$$= \lambda^{k+1} \det A,$$

where $A_1, A_2, \dots, A_k$ are $k \times k$ matrices obtained by expanding across the first row.

By induction, $\det(\lambda A) = \lambda^n \det A$ for an $n \times n$ matrix A.

14. Assume as in the book that $\det(AB) = (\det A)(\det B)$. Then $\det(ABC) = \det[(AB)C] = \det(AB)\det(C) = (\det A)(\det B)(\det C)$. (The student may verify our assumption by brute force.)

17. Multiply the two matrices to get the identity matrix:

$$\begin{bmatrix} a & b \\ c & d \end{bmatrix} \frac{1}{ad-bc} \begin{bmatrix} d & -b \\ -c & a \end{bmatrix} = \frac{1}{ad-bc} \begin{bmatrix} a & b \\ c & d \end{bmatrix} \begin{bmatrix} d & -b \\ -c & a \end{bmatrix} =$$

$$\frac{1}{ad-bc} \begin{bmatrix} ad-bc & 0 \\ 0 & ad-bc \end{bmatrix} = \begin{bmatrix} 1 & 0 \\ 0 & 1 \end{bmatrix}.$$

Similarly, we can show that

$$\left(\frac{1}{ad-bc} \begin{bmatrix} d & -b \\ -c & a \end{bmatrix}\right) \begin{bmatrix} a & b \\ c & d \end{bmatrix} = \begin{bmatrix} 1 & 0 \\ 0 & 1 \end{bmatrix}.$$

1.R Review Exercises for Chapter 1

SOLUTIONS TO SELECTED EXERCISES

1. $\mathbf{v} + \mathbf{w} = (3, 4, 5) + (1, -1, 1) = (4, 3, 6) = 4\mathbf{i} + 3\mathbf{j} + 6\mathbf{k}$; $\mathbf{v} + \mathbf{w}$ is the diagonal of a parallelogram whose sides are $\mathbf{v}$ and $\mathbf{w}$.
   $3\mathbf{v} = 3(3, 4, 5) = (9, 12, 15) = 9\mathbf{i} + 12\mathbf{j} + 15\mathbf{k}$; $3\mathbf{v}$ has the same direction as $\mathbf{v}$ with length 3 times the length of $\mathbf{v}$.
   $6\mathbf{v} + 8\mathbf{w} = 6(3, 4, 5) + 8(1, -1, 1) = (26, 16, 38) = 26\mathbf{i} + 16\mathbf{j} + 38\mathbf{k}$; $6\mathbf{v} + 8\mathbf{w}$ is the diagonal of a parallelogram whose sides are $6\mathbf{v}$ and $8\mathbf{w}$. $6\mathbf{v}$ has the same direction as $\mathbf{v}$ with length 6 times that of $\mathbf{v}$ and similarly for $8\mathbf{w}$.
   $-2\mathbf{v} = -2(3, 4, 5) = (-6, -8, -10) = -6\mathbf{i} - 8\mathbf{j} - 10\mathbf{k}$; $-2\mathbf{v}$ is a vector in the oposite direction of $\mathbf{v}$ with length twice that of $\mathbf{v}$.

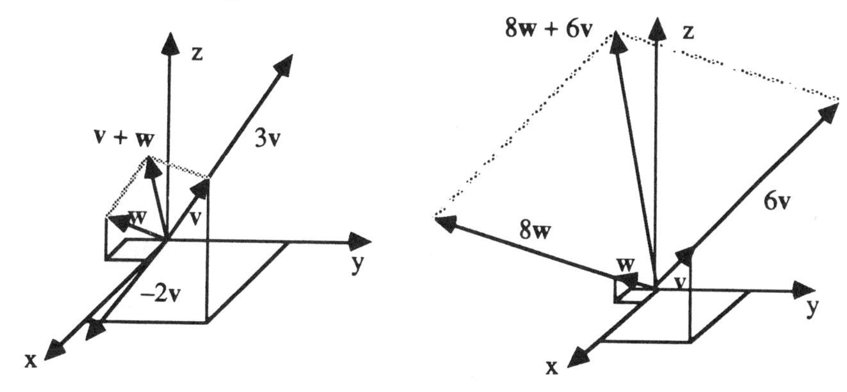

   $\mathbf{v} \cdot \mathbf{w} = (3)(1) + (4)(-1) + (5)(1) = 4$; $\mathbf{v} \cdot \mathbf{w}$ is the number $\|\mathbf{v}\|\,\|\mathbf{w}\| \cos\theta$, where θ is the angle between $\mathbf{v}$ and $\mathbf{w}$.

$$\mathbf{v} \times \mathbf{w} = \begin{vmatrix} \mathbf{i} & \mathbf{j} & \mathbf{k} \\ 3 & 4 & 5 \\ 1 & -1 & 1 \end{vmatrix} = \begin{vmatrix} 4 & 5 \\ -1 & 1 \end{vmatrix}\mathbf{i} - \begin{vmatrix} 3 & 5 \\ 1 & 1 \end{vmatrix}\mathbf{j} + \begin{vmatrix} 3 & 4 \\ 1 & -1 \end{vmatrix}\mathbf{k} = 9\mathbf{i} + 2\mathbf{j} - 7\mathbf{k}$$; $\mathbf{v} \times \mathbf{w}$ is perpendicular

   to both $\mathbf{v}$ and $\mathbf{w}$. Its length is the area of the parallelogram spanned by $\mathbf{v}$ and $\mathbf{w}$.

4. (a) Using the point-direction form of the line, we get $\mathbf{l}(t) = (0, 1, 0) + t(3, 0, 1)$.
(b) Using the point-point form of the line, we get $\mathbf{l}(t) = \mathbf{a} + t(\mathbf{b} - \mathbf{a}) = (0, 1, 1) + t[(0, 1, 0) - (0, 1, 1)] = (0, 1, 1) + t(0, 0, -1)$.
(c) The normal to the plane is $(a, b, c) = (-1, 1, -1)$, so the equation of the plane is
$$a(x - x_0) + b(y - y_0) + c(z - z_0) = 0, \text{ i.e.,}$$
$$-1(x - 1) + 1(y - 1) - 1(z - 1) = 0 \text{ or}$$
$$x - y + z = 1 .$$

5. (b) $\mathbf{v} \cdot \mathbf{w} = (1)(3) + (2)(1) + (-1)(0) = 5$.

6. (b) $$\mathbf{v} \times \mathbf{w} = \begin{vmatrix} \mathbf{i} & \mathbf{j} & \mathbf{k} \\ 1 & 2 & -1 \\ 3 & 1 & 0 \end{vmatrix} = \begin{vmatrix} 2 & -1 \\ 1 & 0 \end{vmatrix}\mathbf{i} - \begin{vmatrix} 1 & -1 \\ 3 & 0 \end{vmatrix}\mathbf{j} + \begin{vmatrix} 1 & 2 \\ 3 & 1 \end{vmatrix}\mathbf{k} = \mathbf{i} - 3\mathbf{j} - 5\mathbf{k} .$$

7. (b) We compute $\|\mathbf{v}\| = \sqrt{1+4+1} = \sqrt{6}$ and $\|\mathbf{w}\| = \sqrt{9+1+0} = \sqrt{10}$. Then the definition of the dot product and the result of exercise 5(b) gives
$$\cos\theta = \frac{\mathbf{v} \cdot \mathbf{w}}{\|\mathbf{v}\|\,\|\mathbf{w}\|} = \frac{5}{\sqrt{6}\sqrt{10}} = \frac{5}{\sqrt{60}} = \frac{5}{2\sqrt{15}} .$$

8. (b) The area of a parallelogram is the length of $\mathbf{v} \times \mathbf{w}$. Using the result of exercise 6(b), we get $\|\mathbf{v} \times \mathbf{w}\| = [1 + 9 + 25]^{1/2} = \sqrt{35}$.

12. We compute the following dot products using the fact that $\mathbf{u}$, $\mathbf{v}$, and $\mathbf{w}$ are orthonormal:
$$\mathbf{a} \cdot \mathbf{u} = (\alpha\,\mathbf{u} + \beta\,\mathbf{v} + \gamma\,\mathbf{w}) \cdot \mathbf{u} = \alpha .$$
$$\mathbf{a} \cdot \mathbf{v} = (\alpha\,\mathbf{u} + \beta\,\mathbf{v} + \gamma\,\mathbf{w}) \cdot \mathbf{u} = \beta .$$
$$\mathbf{a} \cdot \mathbf{w} = (\alpha\,\mathbf{u} + \beta\,\mathbf{v} + \gamma\,\mathbf{w}) \cdot \mathbf{u} = \gamma .$$
Geometrically, $\mathbf{a} \cdot \mathbf{u}$ is the projection of $\mathbf{a}$ on $\mathbf{u}$; similarly for the others.

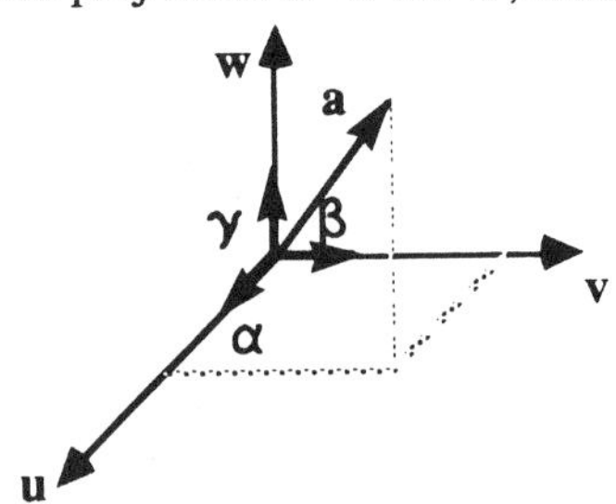

15. From the definition of the dot product and the fact that $\mathbf{u} \cdot \mathbf{u} = \|\mathbf{u}\|^2$, we compute
$$\cos\theta = \frac{\mathbf{v} \cdot \mathbf{a}}{\|\mathbf{v}\|\,\|\mathbf{a}\|} = \frac{\|\mathbf{a}\|\mathbf{b} \cdot \mathbf{a} + \|\mathbf{b}\|\,\|\mathbf{a}\|^2}{\|\mathbf{v}\|\,\|\mathbf{a}\|} = \frac{\mathbf{b} \cdot \mathbf{a} + \|\mathbf{b}\|\,\|\mathbf{a}\|}{\|\mathbf{v}\|} .$$
$$\cos\theta = \frac{\mathbf{v} \cdot \mathbf{b}}{\|\mathbf{v}\|\,\|\mathbf{b}\|} = \frac{\|\mathbf{a}\|\,\|\mathbf{b}\|^2 + \|\mathbf{b}\|\mathbf{a} \cdot \mathbf{b}}{\|\mathbf{v}\|\,\|\mathbf{b}\|} = \frac{\|\mathbf{a}\|\,\|\mathbf{b}\| + \mathbf{b} \cdot \mathbf{a}}{\|\mathbf{v}\|} .$$

Since the angle between $\mathbf{a}$ and $\mathbf{v}$ is the same as the angle between $\mathbf{b}$ and $\mathbf{v}$, the vector $\mathbf{v}$ must bisect the angle between $\mathbf{a}$ and $\mathbf{b}$.

18. (a) The answer is yes: $\mathbf{a} \cdot \mathbf{b} = \mathbf{a}' \cdot \mathbf{b}$ implies $\mathbf{a} \cdot \mathbf{b} - \mathbf{a}' \cdot \mathbf{b} = 0$ or $(\mathbf{a} - \mathbf{a}') \cdot \mathbf{b} = 0$ for all $\mathbf{b}$. Choose $\mathbf{b} = \mathbf{a} - \mathbf{a}'$ to get $\| \mathbf{a} - \mathbf{a}' \|^2 = 0$, so $\mathbf{a} - \mathbf{a}' = \mathbf{0}$, which means that $\mathbf{a} = \mathbf{a}'$.

(b) The answer is yes. If $\mathbf{a} \times \mathbf{b} = \mathbf{a}' \times \mathbf{b}$ for all $\mathbf{b}$, we can conclude that $(\mathbf{a} - \mathbf{a}') \times \mathbf{b} = \mathbf{0}$ for all $\mathbf{b}$. Choose $\mathbf{b}$ to be a unit vector orthogonal to $\mathbf{a} - \mathbf{a}'$. By the definition of the cross product, this implies $\| \mathbf{a} - \mathbf{a}' \| = 0$, so $\mathbf{a} = \mathbf{a}'$.

Note that this problem contains a very subtle point: we are *given* that $\mathbf{a} \times \mathbf{b} = \mathbf{a}' \times \mathbf{b}$ for *all* $\mathbf{b}$; if we want to show that the stated property is not true, we must be able to find *some* $\mathbf{b}$ such that for $\mathbf{a} = \mathbf{a}'$, $\mathbf{a} \times \mathbf{b} \neq \mathbf{a}' \times \mathbf{b}$.

22. (e) Note that the x and y coordinates lie in the third quadrant of the xy plane. The definitions from section 1.4 are used: For cylindrical coordinates, we compute:

$$r = \sqrt{x^2 + y^2} = \sqrt{12 + 4} = 4 .$$

$$\theta = \pi + \tan^{-1}(y/x) = \pi + \tan^{-1}(1/\sqrt{3})$$

$$= \pi + \pi/6 = 7\pi/6 .$$

So the cylindrical coordinates are $(4, 7\pi/6, 3)$.

For spherical coordinates, we compute:

$$\rho = \sqrt{x^2 + y^2 + z^2} = \sqrt{12 + 4 + 9} = 5 .$$

$$\varphi = \cos^{-1}(z/\rho) = \cos^{-1}(4/5) .$$

So the spherical coordinates are $(5, 7\pi/6, \cos^{-1}(4/5))$.

23. (b)

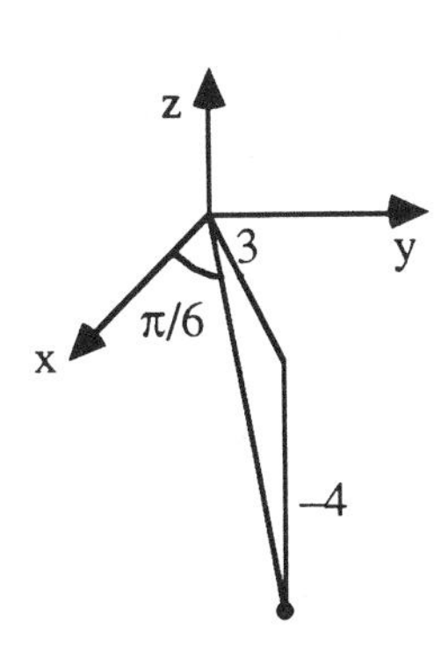

Using the formulas from section 1.4, we calculate:

$$x = r \cos \theta = (3)(\sqrt{3}/2) \text{ and}$$

$$y = r \sin \theta = (3)(1/2) .$$

So the corresponding cartesian coordinates are $(3\sqrt{3}/2, 3/2, -4)$.

For the spherical coordinates, we compute:

$$\rho = \sqrt{x^2 + y^2 + z^2} = \sqrt{\frac{27}{4} + \frac{9}{4} + 16} = 5 .$$

$$\varphi = \cos^{-1}(z/\rho) = \cos^{-1}(-4/5) .$$

So the corresponding spherical coordinates are $(5, \pi/6, \cos^{-1}(-4/5))$.

24. (b)

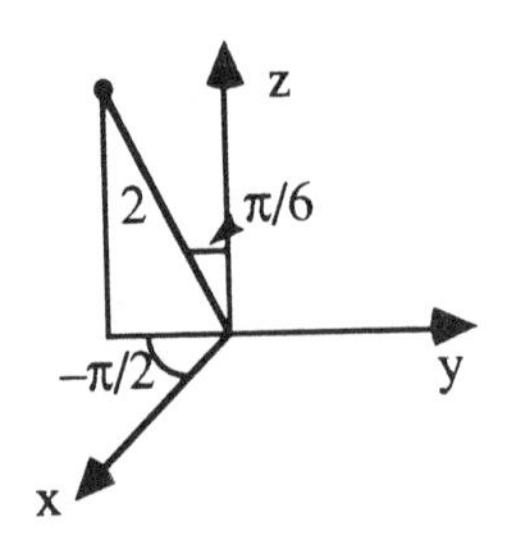

Using the formulas from section 1.4, we compute:

$$x = \rho \cos\theta \sin\varphi = 2(0)(1/2) = 0\ ,$$

$$y = \rho \sin\theta \sin\varphi = 2(-1)(1/2) = -1\ ,$$

$$z = \rho \cos\phi = 2(\sqrt{3}/2) = \sqrt{3}\ .$$

So the corresponding cartesian coordinates are $(0, -1, \sqrt{3})$.

For the cylindrical coordinates, we calculate:

$$r = \sqrt{x^2 + y^2} = \sqrt{0+1} = 1\ .$$

We note that $x = 0$ and $y = -1$, so $\theta = 3\pi/2$. Therefore, the cylindrical coordinates are $(1, 3\pi/2, \sqrt{3})$.

28. Using the methods of section 1.5, we get

$$AB = \begin{bmatrix} 3 & 0 & 1 \\ 2 & 0 & 1 \\ 1 & 0 & 1 \end{bmatrix}\begin{bmatrix} 1 & 0 & 1 \\ 1 & 1 & 1 \\ 0 & 0 & 1 \end{bmatrix} = \begin{bmatrix} 3 & 0 & 4 \\ 2 & 0 & 3 \\ 1 & 0 & 2 \end{bmatrix} \text{ and } BA = \begin{bmatrix} 1 & 0 & 1 \\ 1 & 1 & 1 \\ 0 & 0 & 1 \end{bmatrix}\begin{bmatrix} 3 & 0 & 1 \\ 2 & 0 & 1 \\ 1 & 0 & 1 \end{bmatrix} = \begin{bmatrix} 4 & 0 & 4 \\ 6 & 0 & 3 \\ 1 & 0 & 2 \end{bmatrix}.$$

Clearly, $AB \neq BA$.

33. (a) $\mathbf{r}$ is the vector $7\mathbf{i} + 2\mathbf{j}$, so $W = \mathbf{F} \cdot \mathbf{r} = 70\cos\theta + 20\sin\theta$.

(b) If $\mathbf{F}$ has a magnitude of 6, then $\mathbf{F} = 6\cos\theta\,\mathbf{i} + 6\sin\theta\,\mathbf{j}$. Since $\theta = \pi/6$, we have $\mathbf{F} = 6(\sqrt{3}/2)\mathbf{i} + 6(1/2)\mathbf{j}$, and so $W = \mathbf{F} \cdot \mathbf{r} = (21\sqrt{3} + 6)$ foot-lbs.

36. Subtract the first row from the second and third rows, then expand by minors along the first column to get

$$\begin{vmatrix} 1 & x & x^2 \\ 1 & y & y^2 \\ 1 & z & z^2 \end{vmatrix} = \begin{vmatrix} 1 & x & x^2 \\ 0 & y-x & y^2-x^2 \\ 0 & z-x & z^2-x^2 \end{vmatrix} = \begin{vmatrix} y-x & y^2-x^2 \\ z-x & z^2-x^2 \end{vmatrix} =$$

$$(y-x)(z^2-x^2) - (z-x)(y^2-x^2) = (y-x)(z-x)(z-y) \neq 0$$

if x, y, z are all different. The last step used the fact that $(z^2 - x^2) = (z-x)(z+x)$.

39. (a) We recognize $\mathbf{a} \cdot (\mathbf{b} \times \mathbf{c})$ as a triple product. Let $\mathbf{a} = (a_1, a_2, a_3)$ and use a similar notation for $\mathbf{b}$ and $\mathbf{c}$. Therefore,

$$V = \frac{1}{6}\begin{vmatrix} a_1 & a_2 & a_3 \\ b_1 & b_2 & b_3 \\ c_1 & c_2 & c_3 \end{vmatrix}\ .$$

(b) Use the formula from part (a) and subtract the first row from the second row:

$$\frac{1}{6}\begin{vmatrix} 1 & 1 & 1 \\ 1 & -1 & 1 \\ 1 & 1 & 0 \end{vmatrix} = \frac{1}{6}\begin{vmatrix} 1 & 1 & 1 \\ 0 & -2 & 0 \\ 1 & 1 & 0 \end{vmatrix} = \frac{1}{6}\begin{vmatrix} 1 & 1 \\ 0 & -2 \end{vmatrix} = -1/3\ .$$

We take the absolute value and the volume is 1/3 .

44. A vector which is normal to the first plane is $\mathbf{v} = 8\mathbf{i} + \mathbf{j} + \mathbf{k}$. A vector which is normal to the second plane is $\mathbf{w} = \mathbf{i} - \mathbf{j} - \mathbf{k}$. The cross product $\mathbf{v} \times \mathbf{w}$ is orthogonal to both normal vectors, so it should be parallel to both planes. We compute that $\mathbf{v} \times \mathbf{w} = 2\mathbf{i} - 7\mathbf{j} - 9\mathbf{k}$, so the unit vector in question is $(2\mathbf{i} - 7\mathbf{j} - 9\mathbf{k})/\sqrt{134}$.

47. We want $\mathbf{v} = \alpha\mathbf{i} + \beta\mathbf{j} + \gamma\mathbf{k}$ such that $\|\mathbf{v}\| = 1$. From the definition of the dot product, we know that

$$\cos 30° = \frac{\sqrt{3}}{2} = \frac{\mathbf{v} \cdot \mathbf{i}}{\|\mathbf{v}\|\,\|\mathbf{i}\|}, \text{ so } \mathbf{v} = \frac{\sqrt{3}}{2}\mathbf{i} + \beta\mathbf{j} + \gamma\mathbf{k} .$$

Since $\mathbf{v}$ makes equal angles with $\mathbf{j}$ and $\mathbf{k}$, we must have $\beta = \gamma$. Since $\|\mathbf{v}\| = 1$, we determine that $\beta = \gamma = 1/2\sqrt{2}$. Therefore,

$$\mathbf{v} = \frac{\sqrt{3}}{2}\mathbf{i} + \frac{\sqrt{2}}{4}\mathbf{j} + \frac{\sqrt{2}}{4}\mathbf{k} .$$

CHAPTER 2
DIFFERENTIATION

2.1 The Geometry of Real-Valued Functions

GOALS

1. Be able to define a graph, a level curve, a level set, and a section.

2. Be able to graph a function $f\colon \mathbf{R}^2 \to \mathbf{R}$.

STUDY HINTS

1. *Notation.* $f\colon A \subset \mathbf{R}^n \to \mathbf{R}^m$ describes a function f. The domain is A, which is a subset of $\mathbf{R}^n$. Points in A are mapped to points in the range, which is a subset of $\mathbf{R}^m$. The only restriction on n and m is that they have to be positive integers.

2. *Real-valued function.* In the notation, $f\colon A \subset \mathbf{R}^n \to \mathbf{R}^m$, we restrict m to be 1 in this section. A real-valued function assigns a *unique* real number to each point in A.

3. *Graphs.* The graph of $f\colon A \subset \mathbf{R}^n \to \mathbf{R}$ is drawn in the space $\mathbf{R}^{n+1}$. If $\mathbf{x}$ is a point of A, then the graph consists of all points in $\mathbf{R}^{n+1}$ with the form $(\mathbf{x}, f(\mathbf{x}))$, where $f(\mathbf{x})$ is a real number.

4. *Level set.* This is the graph of f when $f(\mathbf{x})$ is a constant. In $\mathbf{R}^2$, such a set is called a level curve and in $\mathbf{R}^3$, it is called a level surface. Level sets are important for graphing.

5. *Sections.* These are intersections of graphs with a plane. Usually, the most helpful sections for graphs in $\mathbf{R}^3$ are the intersections with the planes $x = \text{constant}$ and $y = \text{constant}$.

6. *Graphing in* $\mathbf{R}^3$. The best way to draw a graph in $\mathbf{R}^3$ is to draw level curves for $z = \text{constant}$. Then lift or drop the curves to the appropriate "height" for $z = \text{constant}$. Analyzing the sections helps complete the graph. It is a good idea to review how to sketch the graph of an ellipse, a hyperbola, a circle, and a parabola from your calculus or precalculus text.

7. *Sketching planes.* Many of us are poor artists, and as a result, three-dimensional geometry may be frustrating due to this problem rather than a lack of mathematical understanding. Planes are most easily sketched by plotting three non-collinear points (usually on the coordinate axes) and then passing a plane through them.

8. *Spheres*. In general, the equation $(x - a)^2 + (y - b)^2 + (z - c)^2 = r^2$ represents a sphere of radius r centered at (a, b, c).

9. *Cylinders*. A surface in $\mathbf{R}^3$ is called a cylinder if x, y, or z is missing from the equation. A cylinder can be sketched by drawing the level curve in the plane where the missing variable is zero. Then move the curve along the axis of the missing variable.

10. *Graphs in* $\mathbf{R}^4$. Example 5 gives a function which can't be drawn on paper. To see the "graph", one can make a movie which shows the concentric spheres expanding.

SOLUTIONS TO SELECTED EXERCISES

1. (a) To determine the level curves, we look at the equation $c = x - y + 2$, where c is a constant. The equation is the same as $y = x + (2 - c)$, which is the equation of a line with slope 1 and y intercept $2 - c$.
The graph of f is a plane with intercepts at $(-2, 0, 0)$, $(0, 2, 0)$, and $(0, 0, 2)$.

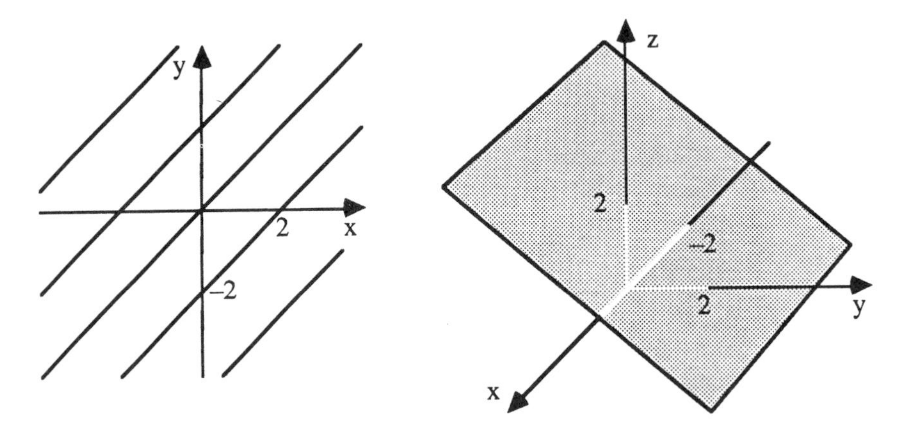

2. (b) We look at $c = 1 - x^2 - y^2$ for $c =$ constant. This rearranges to $x^2 + y^2 = 1 - c$, which is an equation for a circle of radius $\sqrt{1-c}$, centered at the origin if $c < 1$. If $c = 1$, then the level curve is a point at the origin. There is no level curve if $c > 1$.

3. (3) Substitute $x = r \cos \theta$ and $y = r \sin \theta$ to get $z = (r \cos \theta)^2 + (r \sin \theta)^2 = r^2(\cos^2\theta + \sin^2\theta) = r^2$. Since $z = r^2$ does not depend on θ, the shape of the graph does not depend on θ.

5. For constant c, the equation $c = \sqrt{100 - x^2 - y^2}$ is equivalent to $c^2 = 100 - x^2 - y^2$ or $x^2 + y^2 = 100 - c^2$. The level curves are circles of radii $\sqrt{100 - c^2}$, centered at the origin. So, for $c = 0, 2, 4, 6, 8, 10,$ the radii are $10, \sqrt{96}, \sqrt{84}, 8, 6,$ and 0, respectively. Drawing the level curves and raising them to the appropriate z values, we obtain the following graph. The graph of $f(x, y)$ is a hemisphere. The level

curves and the graph are shown here.

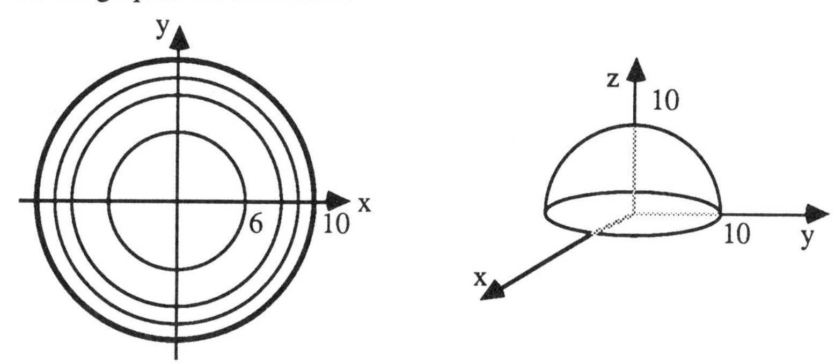

10. The level curves have the equation $c = x/y$ or $y = x/c$ for constant c. These level curves are lines through the origin with slope $1/c$. One restriction is that $y \neq 0$. Notice that when x is held constant, the sections are the hyperbolas $c = yz$, and when y is held constant, we get the lines $c = x/z$. Putting all this information together, we get the "twisted plane" as the graph.

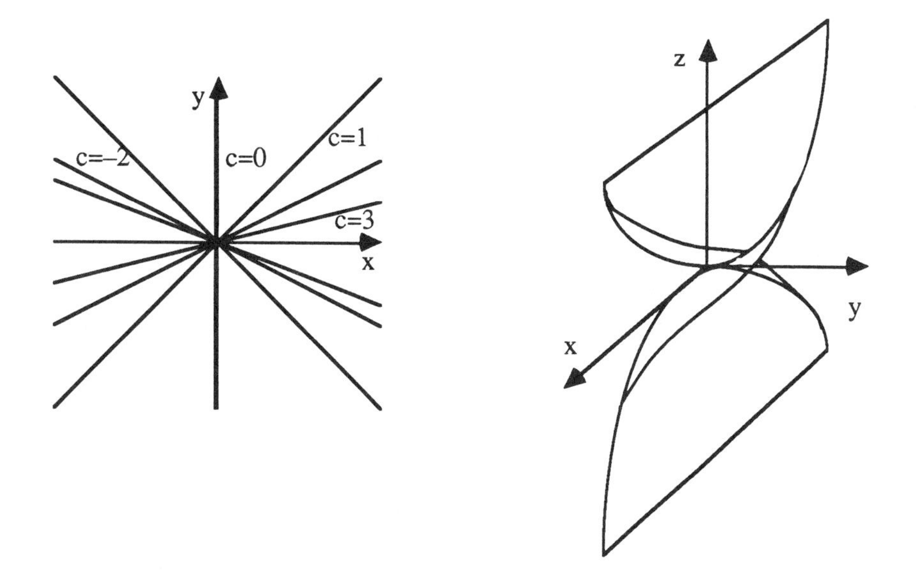

12. The level surfaces have the equation $c = 4x^2 + y^2 + 9z^2$. There is no level surface if $c < 0$. If $c = 0$, then the level surface is the origin. If $c > 0$, then we should look at the level curves for constants k, i.e., analyze $c - 9k^2 = 4x^2 + y^2$. We recognize that if $9k^2 < c$, then the level curves are ellipses which get smaller as $|k|$ approaches $\sqrt{c}/3$. Similarly, we see that level sections parallel to the yz plane have the equation $c - 4k^2 = y^2 + 9z^2$, which are ellipses with decreasing "radii" as $|k|$ approaches $\sqrt{c}/2$. Also, the level sections parallel to the xz plane have the equation $c - k^2 = 4x^2 + 9z^2$,

which are again ellipses which get smaller as $|k|$ approaches $\sqrt{c}$. The level surfaces are ellipsoids if c is positive.

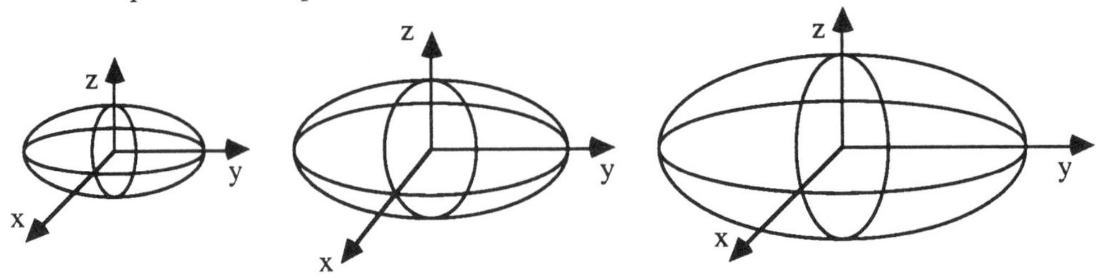

17. In the yz plane, we sketch the graph of $z = |y|$. Since x does not appear in the equation, this sketch is shifted along the x axis to obtain the graph in $\mathbf{R}^3$.

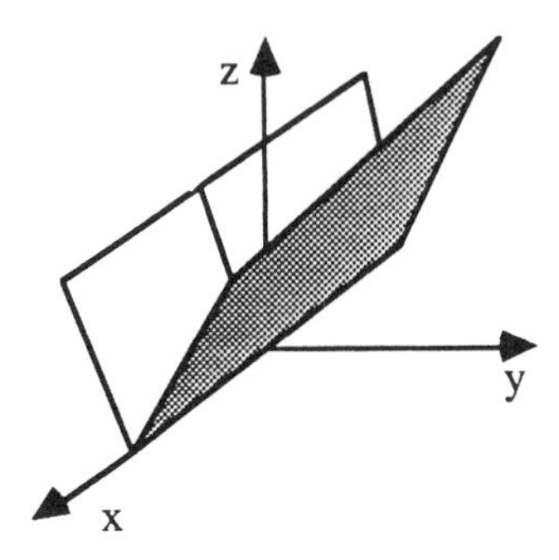

22. The equation can be written as $(x^2 - 2x + 1) + y^2 = 1$ or $(x - 1)^2 + y^2 = 1$. In the xy plane, this is a circle with radius 1, centered at $(1, 0)$. Since z is not in the equation, z may take on any value, so the circle is shifted up and down the z axis.

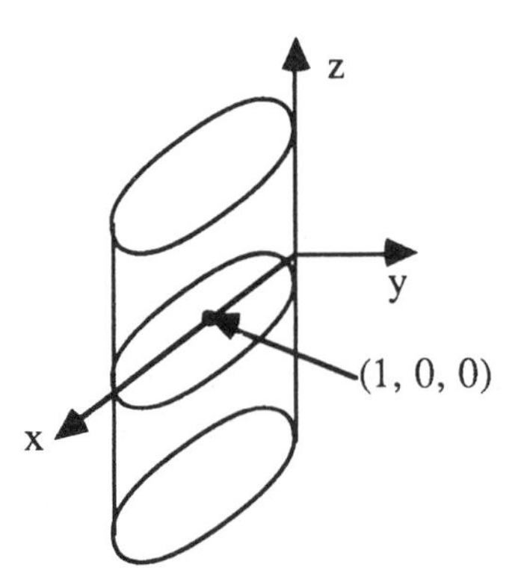

25. Sketch the graph of $z = x^2$ in the xz plane. Then shift the graph along the y axis to get a parabolic cylinder. The graph is shown on the next page.

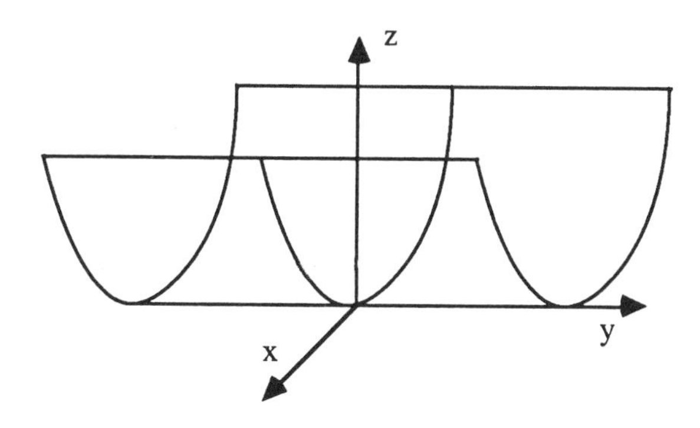

29. An equivalent equation is $4x^2 + 2z^2 = 3y^2$. When $y = 0$, the level curve is the origin. When $y \neq 0$, we have level sections which are ellipses centered around the y axis. The major axes are parallel to the z axis and the minor axes are parallel to the x axis. The ellipses get larger as $|y|$ increases. To complete the graph, we note that when $x = 0$, the section is the straight lines $z = \pm\sqrt{3/2}\ y$. Thus, our graph is a cone.

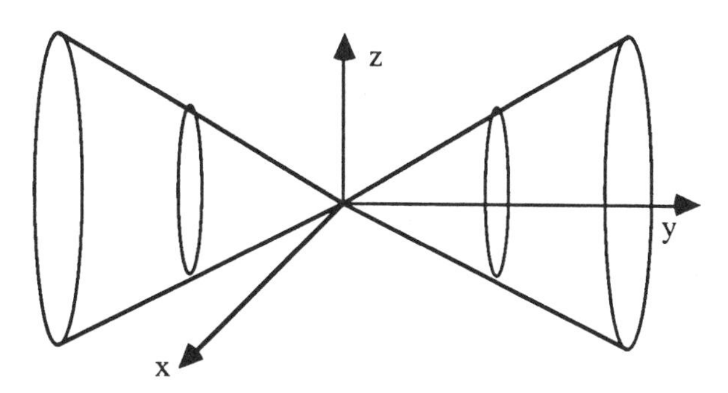

32.

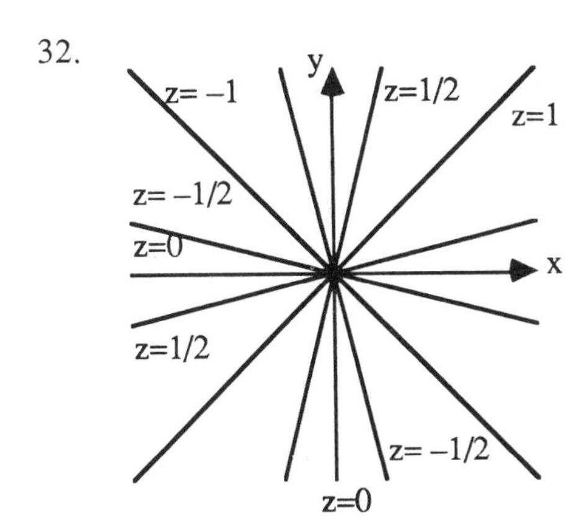

Substitute $x = r\cos\theta$ and $y = r\sin\theta$. Therefore, if $x^2 + y^2 = r^2 \neq 0$, then $z = f(x, y) = 2xy/(x^2 + y^2) = 2(r\cos\theta)(r\sin\theta)/r^2 = r^2(2\sin\theta\cos\theta)/r^2 = \sin 2\theta$. Thus, the function reduces to $z = \sin 2\theta$ if $r \neq 0$. If $-1 \leq z \leq 1$, the level curve is one or two straight lines through the origin satisfying $z = \sin 2\theta$ (see sketch at left). The level curves are raised to a height $z = \sin 2\theta$ to obtain the graph of a "wrinkled plane." (See graph on next page. The dotted line is a portion of the xy plane.) If $z > 1$ or $z < -1$, there is no level curve. Notice that the plane becomes flatter as r gets larger.

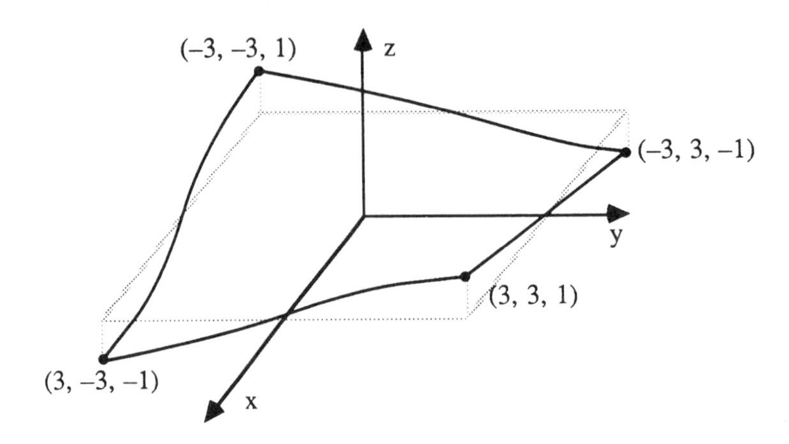

2.2 Limits and Continuity

GOALS

1. Be able to define the following: open disk, open set, neighborhood, boundary point, limit, continuous.

2. Be able to determine where a function is continuous.

3. Given a function, be able to compute a limit or show that a limit does not exist.

STUDY HINTS

1. *Theoretical section.* This section is not essential for computational purposes. Your instructor may choose not to emphasize the material in this section. Use your lectures to determine how important the material is for your course.

2. *Definitions.* (a) An *open disk* is the set of points $\mathbf{x}$ around $\mathbf{x}_0$ such that $\| \mathbf{x} - \mathbf{x}_0 \| < r$. It is denoted $D_r(\mathbf{x}_0)$. Note the strict inequality.
   (b) An *open set* U is a set such that every $\mathbf{x}_0$ has an open disk entirely within U. You will need to find an r when proving that a set is open.
   (c) A *neighborhood* is an open set containing $\mathbf{x}_0$.
   (d) A *boundary point* of a set A has no neighborhood entirely inside or entirely outside of A.

3. *Review.* You should review the concepts of limit and continuity from your one-variable calculus textbook before continuing.

4. *Limits*. In the definition, be aware that $\mathbf{x}_0$ doesn't have to be in A ; $\mathbf{x}_0$ may be on the boundary. Also $f(\mathbf{x}_0)$ doesn't have to be defined. We're only interested in the points $\mathbf{x}$ *near* $\mathbf{x}_0$. For proofs, we need to choose U , which is dependent upon N .

5. *Properties of limits*. Most of these are what you would intuitively expect. Note that for multiplication and division, the mapping is *into* $\mathbf{R}^1$.

6. *Continuity*. Analogous to the one-variable definition, a function is continuous at $\mathbf{x}_0$ if
   $\lim_{\mathbf{x}\to\mathbf{x}_0} f(\mathbf{x}) = f(\mathbf{x}_0)$, i.e., the limit equals the function value.
   The limit on the left-hand side is concerned about points *near* $\mathbf{x}_0$. The right-hand side, $f(\mathbf{x}_0)$, is concerned about the point $\mathbf{x}_0$ itself.

7. *Non-existent limits*. Showing that the limit of f(x, y) does not exist is sometimes simple. To show a limit does not exist, we usually look at the limit of f by first holding x constant then holding y constant. If the two values differ, the limit does not exist.

SOLUTIONS TO SELECTED EXERCISES

1. (b) Take $0 < r \le y$, then for all points (x, y) , the open disk $D_r(x, y) \subset B$.
   (d)

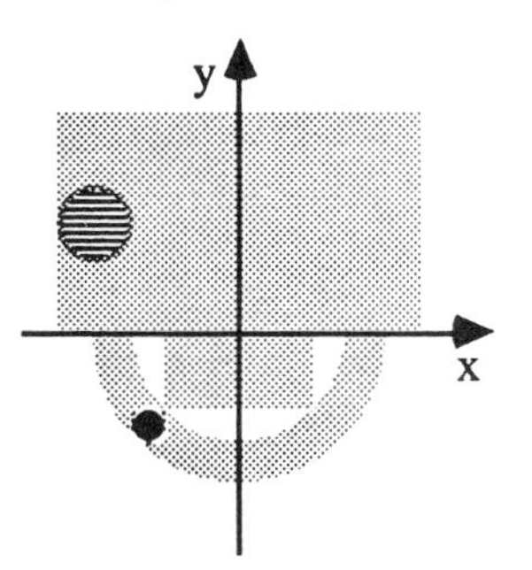

From parts (a), (b), and (c), we know that A, B, and C are open. For $A \cup B \cup C$, take r to be the smallest of those used in parts (a), (b), and (c). In the figure, the striped disk is the same as the one used in part (b) and the black disk is the same as the one used in part (c).

4. (d) Using the properties of limits and the hint in part (c), we compute
$$\lim_{x\to 0} \frac{\sin^2 x}{x^2} = \lim_{x\to 0} \left(\frac{\sin x}{x}\right)^2 = \left(\lim_{x\to 0} \frac{\sin x}{x}\right)^2 = 1^2 = 1 .$$

5. (d) Recall the definition of the derivative: $f'(x_0) = \lim_{h\to 0} \frac{f(x_0+h) - f(x_0)}{h}$. By letting $f(x) = e^x$ and $x_0 = 0$, we get
$$\lim_{h\to 0} \frac{e^h - 1}{h} = \lim_{h\to 0} \frac{e^{0+h} - e^0}{h} = \frac{d}{dx} e^x \Big|_{x=0} = 1 .$$

6. (b) It is obvious that the limit of the numerator is 0 , and the limit of the denominator is $2 \ne 0$, so the limit of the quotient is $0/2 = 0$.

(d) First, hold y constant and let x approach 0. Then use l'Hôpital's rule:

$$\lim_{x \to 0} \frac{\cos x - 1 - (x^2/2)}{x^4} = \lim_{x \to 0} \frac{-\sin x - x}{4x^3} = \lim_{x \to 0} \frac{-\cos x - 1}{12x^2}.$$

The last limit tends to $-\infty$, so the limit does not exist.

7. (c) Use the fact that the limit of a vector is the limit of each component (theorem 3(v)). So we get $\lim_{x \to 1} (x^2, e^x) = (1, e)$.

9. Let $t = xy$, and use theorem 5 to give

$$\lim_{(x, y) \to (0, 0)} \frac{\sin xy}{xy} = \lim_{t \to 0} \frac{\sin t}{t} = 1.$$

12. $f(x) = (1 - x)^8 + \cos(1 + x^3)$ is the sum of two functions. The first is u^8 with $u = 1 - x$. Since u is continuous, theorem 5 says that $(1 - x)^8$ is continuous. The second function is $\cos v$ with $v = 1 + x^3$. Again, since v is continuous, theorem 5 says that $\cos(1 + x^3)$ is continuous. The sum of continuous functions is continuous, so $f(x)$ is continuous.

15. (a) We can make a function continuous by equating $f(\mathbf{x}_0)$ and $\lim_{\mathbf{x} \to \mathbf{x}_0} f(\mathbf{x})$. As in exercise 9, we can let $t = x + y$, so $\lim_{(x, y) \to (0, 0)} \frac{\sin (x + y)}{(x + y)} = \lim_{t \to 0} \frac{\sin t}{t} = 1$. Thus, we let $\frac{\sin (x + y)}{(x + y)} = 1$ for $x = y = 0$, and we have a continuous function.

(b) First, we note that if $x = y$, then $\lim_{(x, y) \to (0, 0)} \frac{xy}{x^2 + y^2} = \lim_{x \to 0} \frac{x^2}{2x^2} = \frac{1}{2}$. On the other hand, if $x = -y$, then $\lim_{(x, y) \to (0, 0)} \frac{xy}{x^2 + y^2} = \lim_{x \to 0} \frac{-x^2}{2x^2} = \frac{-1}{2}$. Since the limiting value depends on the direction of approach, the limit does not exist at the origin, so it is not possible to make the function continuous at the origin.

18. (b)

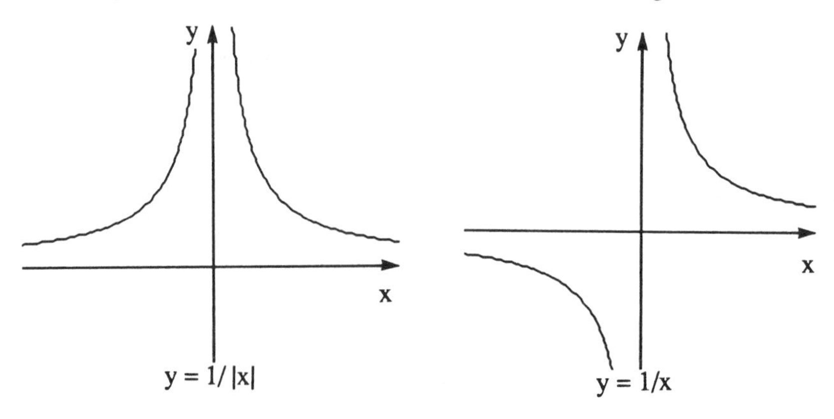

We want to find a δ for every $N > 0$ such that $0 < x < \delta$ implies that $1/|x| >$

N. Let $0 < \delta < 1/N$, then $|x| < 1/N$ implies that $1/|x| > N$. This is not true if the absolute values are omitted, i. e., $\lim_{x \to 0}(1/x)$ may be $+\infty$ or $-\infty$ depending on which side of 0 we are approaching from (see the figure on p. 30).

23. (a) By the triangle inequality, $|a^3 + 3a^2 + a| < |a^3| + 3|a^2| + |a|$; since $|a| < 1$ (we assume it is small, since δ is small), this is less than (or equal to)to $5|a|$. Choose $\delta < (1/500)$, then for $|a| < \delta$, $|a^3 + 3a^2 + a| \leq 1/100$ (note that this is a very rough estimate; a bigger δ would probably work if we work harder to improve the inequality).

2.3 Differentiation

GOALS

1. Be able to state the definition of partial derivatives.
2. Be able to compute a partial derivative or a matrix of partial deerivatives.
3. Be able to compute a gradient.
4. Be able to compute a tangent plane.

STUDY HINTS

1. *Notation.* Class C^n means that the nth derivative is continuous.

2. *Partial derivatives.* Know the definition
$$\frac{\partial f}{\partial x_i} = \lim_{h \to 0} \frac{f(x_1, \ldots, x_i + h, \ldots, x_n) - f(x_1, \ldots, x_i, \ldots, x_n)}{h}.$$
To compute $\partial f/\partial x_i$ without the definition, consider all variables except x_i to be constant and differentiate by one-variable methods. Differentiation is performed with respect to the variable x_i.

3. *Notation for partial derivatives.* In many texts, f_x is used for $\partial f/\partial x$. If we wish to evaluate at a given point, we write
$$\frac{\partial f}{\partial x}(x_0, y_0),\ \left.\frac{\partial f}{\partial x}\right|_{(x_0,y_0)},\ \left.f_x\right|_{(x_0,y_0)},\ \text{or}\ \left.\frac{\partial z}{\partial x}\right|_{(x_0,y_0)}\ \text{if}\ z = f(x, y).$$

4. *Tangent plane.* The tangent plane for a function $f(x, y)$ is given by
$$z = f(x_0, y_0) + \left.\frac{\partial f}{\partial x}\right|_{(x_0, y_0)}(x - x_0) + \left.\frac{\partial f}{\partial y}\right|_{(x_0, y_0)}(y - y_0).$$
This equation is also used to compute a linear approximation. Compare this equation to

the equation of a tangent line and the linear approximation in the one-variable case. See section 2.5 for a generalization.

5. *Differentiability*. Equation (2) tells you that $f: \mathbf{R}^2 \to \mathbf{R}$ is differentiable if the tangent plane approaches $f(x_0, y_0)$ as (x, y) approaches (x_0, y_0). Now, if $f: U \subset \mathbf{R}^n \to \mathbf{R}^m$, then
$$\lim_{\mathbf{x} \to \mathbf{x}_0} \frac{\| f(\mathbf{x}) - f(\mathbf{x}_0) - \mathbf{T}(\mathbf{x} - \mathbf{x}_0) \|}{\| \mathbf{x} - \mathbf{x}_0 \|} = 0,$$
where $\mathbf{T}: \mathbf{R}^n \to \mathbf{R}^m$ is the derivative. You should be able to get equation (2) from this definition. This definition of differentiability is most important for theoretical work.

6. *Gradient.* The gradient is a vector whose components are the partial derivatives of f, with $\partial f/\partial x_i$ in the ith position. Here, f is a real-valued function. This operation is denoted by the symbol ∇. Sometimes, it is denoted "grad." For a function $f: \mathbf{R}^3 \to \mathbf{R}$, you should remember the formula
$$\nabla f = \frac{\partial f}{\partial x}\mathbf{i} + \frac{\partial f}{\partial y}\mathbf{j} + \frac{\partial f}{\partial z}\mathbf{k} .$$

7. *Derivative of vector-valued functions.* Consider a function $f: \mathbf{R}^n \to \mathbf{R}^m$. The derivative is a $m \times n$ matrix of partial derivatives. The range consists of vectors with m components. Think of the components as real-valued vector functions, then each row of the derivative matrix is a gradient. The derivative matrix of f, evaluated at $\mathbf{x}_0$, is denoted $\mathbf{D}f(\mathbf{x}_0)$.

8. *Important facts.* Differentiablity implies continuity of the function, but continuity does not imply differentiability. The existence of continuous partial derivatives implies differentiability but the converse is not true. If a function is differentiable, then its partial derivatives exist, but the converse is, again, not true.

SOLUTIONS TO SELECTED EXERCISES

1. (b) Holding y constant and differentiating with respect to x, we get $\partial f/\partial x = ye^{xy}$. By symmetry, $\partial f/\partial y = xe^{xy}$. In this problem, all we did to compute $\partial f/\partial y$ was to switch x and y. This is what we mean by "symmetry."

2. (b) Hold y constant and use the chain rule to differentiate with respect to x. We get
$$\frac{\partial z}{\partial x} = \frac{1}{\sqrt{1+xy}} \bullet \frac{1}{2} \frac{1}{\sqrt{1+xy}} \bullet y = \frac{y}{2(1+xy)}\text{; similarly, } \frac{\partial z}{\partial y} = \frac{x}{2(1+xy)} .$$
At $(1, 2)$, $\partial z/\partial x = 1/3$ and $\partial z/\partial y = 1/6$.
At $(0, 0)$, $\partial z/\partial x = \partial z/\partial y = 0$.

3. (b) Holding y constant and using the quotient rule, we calculate:
$$\frac{\partial w}{\partial x} = \frac{2x(x^2 - y^2) - 2x(x^2 + y^2)}{(x^2 - y^2)^2} = \frac{-4xy^2}{(x^2 - y^2)^2} .$$

Holding x constant and differentiating with the quotient rule, we get

$$\frac{\partial w}{\partial y} = \frac{2y(x^2 - y^2) - (-2y)(x^2 + y^2)}{(x^2 - y^2)^2} = \frac{4x^2y}{(x^2 - y^2)^2}.$$

4. (b) We must show that the partials are continuous in the domain: $\partial f/\partial x = 1/y - y/x^2 = (x^2 - y^2)/x^2y$, which is continuous for $x \neq 0$ and $y \neq 0$; $\partial f/\partial y = -x/y^2 + 1/x = (y^2 - x^2)/xy^2$, which is continuous for $x \neq 0$ and $y \neq 0$. f(x) is C^1 since it is continuously differentiable.

6. (b) The equation of the tangent plane is given by $z = z_0 + [f_x(x_0, y_0)](x - x_0) + [f_y(x_0, y_0)](y - y_0)$. Using the result of exercise 1(b), we compute:

$$\frac{\partial f}{\partial x}\Big|_{(0,1)} = 1; \quad \frac{\partial f}{\partial y}\Big|_{(0,1)} = 0; \quad f(0, 1) = 1.$$

Therefore, the tangent plane is $z = 1 + 1(x - 0) + 0(y - 1)$ or $z = 1 + x$.

7. (b) The first row contains the partial derivatives of $xe^y + \cos y$. The second row contains those of x, and the third row contains those of $x + e^y$. The first column contains the partial derivatives with respect to x and the second column contains those with respect to y. Thus, the matrix of partial derivatives is

$$\begin{bmatrix} e^y & xe^y - \sin y \\ 1 & 0 \\ 1 & e^y \end{bmatrix}.$$

8. (b) The function f is a mapping from $\mathbf{R}^3$ to $\mathbf{R}^2$, so the matrix of the partials is 2×3. Let $f_1(x, y, z) = x - y$, the first component of f. Similarly, let $f_2(x, y, z) = y + z$. Then

$$\mathbf{D}f(x, y, z) = \begin{bmatrix} \frac{\partial f_1}{\partial x} & \frac{\partial f_1}{\partial y} & \frac{\partial f_1}{\partial z} \\ \frac{\partial f_2}{\partial x} & \frac{\partial f_2}{\partial y} & \frac{\partial f_2}{\partial z} \end{bmatrix} = \begin{bmatrix} 1 & -1 & 0 \\ 0 & 1 & 1 \end{bmatrix}.$$

11. Using the result of exercise 1(b), $x(\partial f/\partial x) = xy\, e^{xy} = y(\partial f/\partial y)$.

12. (b) Use the linear approximation formula, which is the same as the equation of the tangent plane: $z = z_0 + [f_x(x_0, y_0)](x - x_0) + [f_y(x_0, y_0)](y - y_0)$. Let $z = f(x, y) = x^3 + y^3 - 6xy$, $x_0 = 1$, $x = 0.99$, $y_0 = 2$, and $y = 2.01$. We compute:

$\partial f/\partial x = 3x^2 - 6y$, so $f_x(1, 2) = -9$.

$\partial f/\partial y = 3y^2 - 6x$, so $f_y(1, 2) = 6$.

Therefore, our linear approximation is $z \approx -3 + (-9)(-0.01) + (6)(0.01) = -2.85$. The actual value is -2.8485.

13. (c) The gradient is defined as the vector $(\partial f/\partial x, \partial f/\partial y, \partial f/\partial z)$. So $\nabla f(x, y, z) = z^2 e^x \cos y\, \mathbf{i} - z^2 e^x \sin y\, \mathbf{j} + 2z\, e^x \cos y\, \mathbf{k}$.

14. (c) The tangent plane is defined by $\nabla f(\mathbf{x}_0) \cdot (\mathbf{x} - \mathbf{x}_0) = 0$. From exercise 13(c), we have $\nabla f(1, 0, 1) = e\mathbf{i} + 2e\mathbf{k}$, so the tangent plane is $e(x-1) + 2e(z-1) = 0$ or $x + 2z = 3$.

17. We compute $\nabla f(0, 0, 1) = (2x, 2y, -2z)\,|_{(0,0,1)} = (0, 0, -2) = -2\mathbf{k}$.

20. We want to find $\mathbf{T}$ in equation (4). By linearity, $f(\mathbf{x}) - f(\mathbf{x}_0) = f(\mathbf{x} - \mathbf{x}_0)$. Denoting $\mathbf{x} - \mathbf{x}_0$ by $\mathbf{h}$, thus want to find $\mathbf{T}$ so that

$$\lim_{\mathbf{h} \to \mathbf{0}} \frac{\| f(\mathbf{h}) - \mathbf{T}\mathbf{h} \|}{\| \mathbf{h} \|} = 0.$$

If we choose $\mathbf{T} = f$, the numerator vanishes for all $\mathbf{h}$, so this $\mathbf{T}$ satisfies the condition, or the derivative of a linear map is the map itself. For example, in one variable, consider $f(x) = ax$. From one-variable calculus, $\mathbf{T} = f'(x_0) = a$ for all x_0.

2.4 Properties of the Derivative

GOALS

1. Be able to state the chain rule.

2. Be able to compute a partial derivative by using the chain rule.

STUDY HINTS

1. *Chain rule.* Suppose f is a function of $y_1, y_2, \ldots, y_n$ and each y_i is a function of x, then

$$\frac{df}{dx} = \frac{\partial f}{\partial y_1} \cdot \frac{dy_1}{dx} + \frac{\partial f}{\partial y_2} \cdot \frac{dy_2}{dx} + \ldots + \frac{\partial f}{\partial y_n} \cdot \frac{dy_n}{dx}.$$

Notice how each term appears to be df/dx with ∂y_i and dy_i "cancelling". However, beware that the "sum" on the right-hand side is df/dx, *not* n times df/dx. Also note the different d's : "∂" is for a function of many variables, while "d" is for a function of *one* variable.

2. *Multivariable chain rule.* The multivariable chain rule states that

$$\mathbf{D}(f \circ g)(\mathbf{x}_0) = \mathbf{D}f(\mathbf{y}_0)\,\mathbf{D}g(\mathbf{x}_0) \quad \text{where } \mathbf{y} = g(\mathbf{x}_0).$$

This is the product of two derivative matrices, so any desired partial may be obtained by multiplication.

3. *Chain rule, gradient relationship.* Know that if f is real-valued and $h(t) = f(\mathbf{c}(t))$, then

$$\frac{dh}{dt} = \nabla f(\mathbf{c}(t)) \cdot \mathbf{c}'(t) .$$

SOLUTIONS TO SELECTED EXERCISES

2. (b) The function f is differentiable by the sum rule. Its derivative is
$$\left[\frac{\partial f}{\partial x}, \frac{\partial f}{\partial y}\right] = [1 , 1] .$$
(f) The function is differentiable by the chain rule. We know that x^2 and y^2 are differentiable by the product rule and that $1 - x^2 - y^2$ is differentiable by the sum rule. Also, the square root function is differentiable, so the entire function is differentiable. Its derivative is
$$\left[\frac{\partial f}{\partial x}, \frac{\partial f}{\partial y}\right] = \left[\frac{-x}{\sqrt{1 - x^2 - y^2}}, \frac{-y}{\sqrt{1 - x^2 - y^2}}\right] .$$

3. (b) This is a special case of the first special case of the chain rule:
$$\frac{dh}{dx} = \frac{\partial f}{\partial x} \cdot \frac{dx}{dx} + \frac{\partial f}{\partial u} \cdot \frac{du}{dx} + \frac{\partial f}{\partial v} \cdot \frac{dv}{dx}$$
$$= \frac{\partial f}{\partial x} + \frac{\partial f}{\partial u} \cdot \frac{du}{dx} + \frac{\partial f}{\partial v} \cdot \frac{dv}{dx} .$$

5. (b) First, we compute $f(\mathbf{c}(t)) = \exp(3t^2 \cdot t^3) = \exp(3t^5)$, so
$$f'(t) = 15t^4 \exp(3t^5) .$$
Then, by the chain rule,
$$\frac{df}{dt} = \frac{\partial f}{\partial x} \cdot \frac{dx}{dt} + \frac{\partial f}{\partial y} \cdot \frac{dy}{dt} = y\, e^{xy} \cdot 6t + x\, e^{xy} \cdot 3t^2 .$$
Now, substitute $x = 3t^2$ and $y = t^3$ to get
$$df/dt = t^3 \exp(3t^5) \cdot 6t + 3t^2 \exp(3t^2) \cdot 3t^2$$
$$= 15t^4 \exp(3t^5) ,$$
which is the same as we got from a direct computation.

6. (b) Take the derivative of each component to get $\mathbf{c}'(t) = (6t, 3t^2)$.

9. Substitute $u = e^{x-y}$ and $v = x - y$ to get
$$f \circ g = (\tan(e^{x-y} - 1) - e^{x-y}, (e^{x-y})^2 - (x - y)^2) .$$
By the chain rule,
$$\mathbf{D}(f \circ g)(x, y) = \mathbf{D}f(u, v) \cdot \mathbf{D}g(x, y) .$$
First, we calculate
$$\mathbf{D}f(u, v) = \begin{bmatrix} \partial f_1/\partial u & \partial f_1/\partial v \\ \partial f_2/\partial u & \partial f_2/\partial v \end{bmatrix} = \begin{bmatrix} \sec^2(u-1) & -e^v \\ 2u & -2v \end{bmatrix} .$$
When $(x, y) = (1, 1)$, we have $g(1, 1) = (e^{1-1}, 1 - 1) = (1, 0)$. Hence,
$$\mathbf{D}f(1, 0) = \begin{bmatrix} 1 & -1 \\ 2 & 0 \end{bmatrix} .$$
Next, we calculate

$$\mathbf{D}g(x, y) = \begin{bmatrix} \partial g_1/\partial x & \partial g_1/\partial y \\ \partial g_2/\partial x & \partial g_2/\partial y \end{bmatrix} = \begin{bmatrix} e^{x-y} & -e^{x-y} \\ 1 & -1 \end{bmatrix}; \text{ so } \mathbf{D}g(1, 1) = \begin{bmatrix} 1 & -1 \\ 1 & -1 \end{bmatrix}.$$

Therefore $\mathbf{D}(f \circ g)(1, 1) = \begin{bmatrix} 1 & -1 \\ 2 & 0 \end{bmatrix}\begin{bmatrix} 1 & -1 \\ 1 & -1 \end{bmatrix} = \begin{bmatrix} 0 & 0 \\ 2 & -2 \end{bmatrix}$.

Alternatively, we may calculate $\mathbf{D}(f \circ g)(x, y)$ directly from $(f \circ g)(x, y)$.

13. (a) By the chain rule,

$$\frac{dT}{dt} = \nabla T(\mathbf{c}(t)) \cdot \mathbf{c}'(t), \text{ where}$$

$$\mathbf{c}(t) = (\cos t, \sin t).$$

Differentiate:

$$\nabla T(x, y) = \left(\frac{\partial T}{\partial x}, \frac{\partial T}{\partial y}\right) = (2xe^y - y^3, x^2e^y - 3xy^2).$$

Substituting $x = \cos t$, $y = \sin t$ gives

$$\nabla T(\mathbf{c}(t)) = (2 \cos t\, e^{\sin t} - \sin^3 t, \cos^2 t\, e^{\sin t} - 3 \cos t \sin^2 t),$$

$$\mathbf{c}'(t) = (-\sin t, \cos t).$$

Thus,

$$\frac{dT}{dt} = -2 \sin t \cos t\, e^{\sin t} + \sin^4 t + \cos^3 t\, e^{\sin t} - 3 \cos^2 t \sin^2 t.$$

(b) Plug in $x = \cos t$, $y = \sin t$ into the expression for T and get

$$T(t) = \cos^2 t\, e^{\sin t} - \cos t \sin^3 t.$$

Using techniques from one-variable calculus,

$$\frac{dT}{dt} = -2 \sin t \cos t\, e^{\sin t} + \cos^3 t\, e^{\sin t} + \sin^4 t - 3 \cos^2 t \sin^2 t,$$

which is the same as the answer obtained by the chain rule in part (a).

17. (a) If $y(x)$ and G are differentiable, then we can write

$$\frac{dG}{dx} = \frac{\partial G}{\partial x} \cdot \frac{dx}{dx} + \frac{\partial G}{\partial y} \cdot \frac{dy}{dx} = \frac{\partial G}{\partial x} + \frac{\partial G}{\partial y} \cdot \frac{dy}{dx} = 0.$$

Solve for dy/dx : If $\partial G/\partial y \neq 0$, then

$$\frac{dy}{dx} = -\frac{\partial G/\partial x}{\partial G/\partial y}.$$

(b) As in part (a), we differentiate G_1 and G_2 by the chain rule:

$$\frac{dG_1}{dx} = \frac{\partial G_1}{\partial x} + \frac{\partial G_1}{\partial y_1} \cdot \frac{dy_1}{dx} + \frac{\partial G_1}{\partial y_2} \cdot \frac{dy_2}{dx} = 0 \quad \text{and}$$

$$\frac{dG_2}{dx} = \frac{\partial G_2}{\partial x} + \frac{\partial G_2}{\partial y_1} \cdot \frac{dy_1}{dx} + \frac{\partial G_2}{\partial y_2} \cdot \frac{dy_2}{dx} = 0.$$

Assuming $y_1(x)$, $y_2(x)$, and G are differentiable and

$$\begin{vmatrix} \partial G_1/\partial y_1 & \partial G_1/\partial y_2 \\ \partial G_2/\partial y_1 & \partial G_2/\partial y_2 \end{vmatrix} \neq 0 \text{ for all } x,$$

then we can solve for dy_1/dx and dy_2/dx . Rewrite the two equations as

$$\frac{\partial G_1}{\partial y_1} \cdot \frac{dy_1}{dx} + \frac{\partial G_2}{\partial y_2} \cdot \frac{dy_2}{dx} = -\frac{\partial G_1}{\partial x} \qquad (1)$$

$$\frac{\partial G_1}{\partial y_1} \bullet \frac{dy_1}{dx} + \frac{\partial G_2}{\partial y_2} \bullet \frac{dy_2}{dx} = -\frac{\partial G_1}{\partial x} \qquad (1)$$

$$\frac{\partial G_2}{\partial y_1} \bullet \frac{dy_1}{dx} + \frac{\partial G_2}{\partial y_2} \bullet \frac{dy_2}{dx} = -\frac{\partial G_2}{\partial x} \qquad (2)$$

Multiply (1) by $\partial G_2/\partial y_1$ and (2) by $-\partial G_1/\partial y_1$. Add the two together to get:

$$\frac{dy_2}{dx} = \frac{-\frac{\partial G_1}{\partial x} \bullet \frac{\partial G_2}{\partial y_1} + \frac{\partial G_2}{\partial x} \bullet \frac{\partial G_1}{\partial y_1}}{\frac{\partial G_1}{\partial y_2} \bullet \frac{\partial G_2}{\partial y_1} - \frac{\partial G_2}{\partial y_2} \bullet \frac{\partial G_1}{\partial y_1}} .$$

Similarly,

$$\frac{dy_1}{dx} = \frac{-\frac{\partial G_1}{\partial x} \bullet \frac{\partial G_2}{\partial y_2} + \frac{\partial G_2}{\partial x} \bullet \frac{\partial G_1}{\partial y_2}}{\frac{\partial G_1}{\partial y_1} \bullet \frac{\partial G_2}{\partial y_2} - \frac{\partial G_2}{\partial y_1} \bullet \frac{\partial G_1}{\partial y_2}} .$$

18. Begin with $F(x, y, z) = 0$. Let $x = f(y, z)$, $y = g(x, z)$, and $z = h(x, y)$; This means that $F(f(y, z), y, z) = 0$, $F(x, g(x, z), z) = 0$, and $F(x, y, h(x, y)) = 0$. Differentiate $F(f(y, z), y, z)$ with respect to y and z, we get

$$F_x\frac{\partial x}{\partial y} + F_y = 0, \qquad (1) \qquad \text{and}$$

$$F_x\frac{\partial x}{\partial z} + F_z = 0. \qquad (2)$$

Similarly, differentiating $F(x, g(x, z), z)$ and $F(x, y, h(x, y))$, with respect to x and z and y and z, respectively, we get

$$F_x + F_y\frac{\partial y}{\partial x} = 0, \qquad (3)$$

$$F_y\frac{\partial y}{\partial z} + F_z = 0, \qquad (4) \qquad \text{and}$$

$$F_x + F_z\frac{\partial z}{\partial x} = 0, \qquad (5)$$

$$F_y + F_z\frac{\partial z}{\partial y} = 0. \qquad (6)$$

Solving (2) for $\partial x/\partial z$, (3) for $\partial y/\partial x$, and (6) for $\partial z/\partial y$ gives

$$\frac{\partial x}{\partial z} = -\frac{F_z}{F_x}, \quad \frac{\partial y}{\partial x} = -\frac{F_x}{F_y}, \text{ and } \frac{\partial z}{\partial y} = -\frac{F_y}{F_z},$$

assuming that none of the partials F_x, F_y, and F_z are 0. Multiply and get -1.

20. (a) Use the definition of the partial derivative:

$$\frac{\partial f}{\partial x}(0, 0) = \lim_{h \to 0} \frac{f(0 + h, 0) - f(0, 0)}{h} = \lim_{h \to 0} \frac{\frac{h(0)^2}{h^2 + 0^2} - 0}{h} = \lim_{h \to 0} \frac{0}{h} = 0 .$$

The last step holds since $0/h = 0$ for all $h \neq 0$. Similarly,

$$\frac{\partial f}{\partial y}(0,0) = \lim_{h \to 0} \frac{f(0, 0+h) - f(0,0)}{h} = \lim_{h \to 0} \frac{\frac{0(h^2)}{0^2 + h^2} - 0}{h} = \lim_{h \to 0} \frac{0}{h} = 0 .$$

Therefore, $\partial f/\partial x$ and $\partial f/\partial y$ exist at $(0, 0)$ and equals 0.

(b) If $\mathbf{g}(t) = (at, bt)$, then

$$(f \circ g)(t) = \frac{(at)(bt)^2}{(at)^2 + (bt)^2} = \frac{ab^2t^3}{(a^2 + b^2)t^2} = \frac{ab^2}{a^2 + b^2} t ,$$

so

$$(f \circ g)'(t) = \frac{ab^2}{a^2 + b^2} .$$

On the other hand, from part (a), we have $\nabla f(0, 0) = (\partial f/\partial x, \partial f/\partial y)(0, 0) = (0, 0)$. Also, we compute $\mathbf{g}'(t) = (a, b)$, so $\nabla f(0, 0) \cdot \mathbf{g}'(0) = (0, 0) \cdot (a, b) = 0$. Therefore, the chain rule does not apply to this function.

24. $\partial w/\partial x$ on the left hand side means the partial derivative of $w(x, y, g(x, y))$ with respect to x, holding y constant; $\partial w/\partial x$ on the right hand side means the partial derivative of $w(x, y, z)$ with respect to x, holding y and z constant. So the right hand side is not equal to the left hand side.

2.5 Gradients and Directional Derivatives

GOALS

1. Be able to define a directional derivative.
2. Be able to compute a directional derivative.
3. Be able to explain the significance of the gradient.
4. Be able to understand the relationships among the directional derivative, the gradient, the tangent plane, and level sets.

STUDY HINTS

1. *Important example 1.* Many examples in the book use the fact that $\nabla r = \mathbf{r}/r$ where $\mathbf{r} = (x, y, z)$ and $r = \|\mathbf{r}\| = \sqrt{x^2 + y^2 + z^2}$. This is derived in example 1. Much time can be saved by remembering this result.

2. *Definition.* The directional derivative is defined to be

$$\frac{d}{dt} f(\mathbf{x} + t\mathbf{v}) \Big|_{t=0} \text{ or } \lim_{h \to 0} \frac{f(\mathbf{x} + h\mathbf{v}) - f(\mathbf{x})}{h} .$$

The directional derivative gives a rate of change in the direction of $\mathbf{v}$.

3. *Geometric interpretation.* Suppose $\mathbf{v} = a\mathbf{i} + b\mathbf{j}$ is a unit vector, (x_0, y_0) is a given point, and $f(x, y)$ is a surface. The directional derivative tells us the "slope" of a curve at (x_0, y_0) in the direction of $\mathbf{v}$. The curve is formed by the intersection of the surface with the plane described by the set of points $s\mathbf{v} + t\mathbf{k}$. If $\mathbf{v}$ is not a unit vector, then the "slope" may be determined by normalizing $\mathbf{v}$ to be a unit vector.

4. *Relation to partial derivatives.* The partials $\partial f/\partial x$, $\partial f/\partial y$, and $\partial f/\partial z$ are the directiona derivatives in the directions of $\mathbf{i}, \mathbf{j}$, and $\mathbf{k}$, respectively.

5. *Computing directional derivatives.* To compute the directional derivative of f in the direction of $\mathbf{v}$, the easiest formula to use is $\nabla f(\mathbf{x}) \cdot \mathbf{v}$. The directional derivative is a scalar, *not* a vector.

6. *Gradient properties.* Recall that $\nabla f = (\partial f/\partial x)\mathbf{i} + (\partial f/\partial y)\mathbf{j} + (\partial f/\partial z)\mathbf{k}$. You should know that ∇f points in the direction in which f is increasing fastest and $-\nabla f$ points in the direction of fastest decrease. The gradient is always orthogonal to a level surface of f .

7. *Tangent plane.* In terms of the gradient, the tangent plane is $\nabla f(\mathbf{x}_0) \cdot (\mathbf{x} - \mathbf{x}_0) = 0$. This generalizes the formula given in section 2.3.

SOLUTIONS TO SELECTED EXERCISES

1. The directional derivative of $f(\mathbf{x})$ at $\mathbf{x}_0$ in the direction $\mathbf{d}$ is $\nabla f(\mathbf{x}_0) \cdot \mathbf{d}$.

$$\begin{aligned} &\nabla f(1, 1, 2) \cdot (1/\sqrt{5}, 2/\sqrt{5}, 0) \\ &= (z^2, 3y^2, 2xz)\,|_{(1,1,2)} \cdot (1/\sqrt{5}, 2/\sqrt{5}, 0) \\ &= (4, 3, 4) \cdot (1/\sqrt{5}, 2/\sqrt{5}, 0) \\ &= 10/\sqrt{5} = 2\sqrt{5}\,. \end{aligned}$$

2. (b) Given $f(x, y) = \ln(\sqrt{x^2 + y^2})$, we compute

$$\nabla f(x, y) = \left(\frac{1}{\sqrt{x^2+y^2}} \cdot \frac{2x}{2\sqrt{x^2+y^2}}\right)\mathbf{i} + \left(\frac{1}{\sqrt{x^2+y^2}} \cdot \frac{2y}{2\sqrt{x^2+y^2}}\right)\mathbf{j}$$

$$= \frac{x}{x^2+y^2}\mathbf{i} + \frac{y}{x^2+y^2}\mathbf{j}\,.$$

So $\nabla f(1, 0) = (1, 0)$ and the directional derivative is $\nabla f(1, 0) \cdot \mathbf{v} = 2/\sqrt{5}$.

3. (b) Given $f(x, y, z) = e^x + yz$, we calculate $\nabla f(x, y, z) = e^x\mathbf{i} + z\mathbf{j} + y\mathbf{k}$ and $\nabla f(1, 1, 1) = e\mathbf{i} + \mathbf{j} + \mathbf{k}$. The unit vector parallel to $(1, -1, 1)$ is $(1, -1, 1)/\sqrt{3}$.

Thus, the directional derivative is $\nabla f(1, 1, 1) \cdot (1, -1, 1)/\sqrt{3} = e/\sqrt{3}$.

4. (c) For a function of three variables, (x, y, z) , the tangent plane to the surface is $\nabla f(x_0, y_0, z_0) \cdot (x - x_0, y - y_0, z - z_0) = 0$. To use this formula, we need to describe the surface by $f(x, y, z) = \text{constant}$. In this case, $f(x, y, z) = xyz = 1$, so $\nabla f(x, y, z) = (yz, xz, xy)$ and at $(1, 1, 1)$, $\nabla f = (1, 1, 1)$. Therefore, the desired tangent plane is $(1, 1, 1) \cdot (x - 1, y - 1, z - 1) = 0$ or $x + y + z = 3$.

5. (b) $z = (\cos x)(\cos y)$, so $z_x = -\sin x \cos y$ and by symmetry, $z_y = -\sin y \cos x$. At $(0, \pi/2, 0)$, we compute that $z_x = 0$ and $z_y = -1$. So the equation of the tangent plane is $z = z_0 + [z_x(x_0, y_0)](x - x_0) + [z_y(x_0, y_0)](y - y_0) = 0 + 0(x - 0) - 1(y - \pi/2)$ or $z + y = \pi/2$.

6. (c) Given $f(x, y, z) = 1/(x^2 + y^2 + z^2) = 1/r^2$, we have $f_x = -2x/(x^2 + y^2 + z^2)^2 = -2x/r^4$. By symmetry, $f_y = -2y/r^4$ and $f_z = -2z/r^4$. Then $\nabla f = (-2/r^4)(x\mathbf{i} + y\mathbf{j} + z\mathbf{k}) = -2\mathbf{r}/r^4$, where $\mathbf{r} = x\mathbf{i} + y\mathbf{j} + z\mathbf{k}$ and $r = \sqrt{x^2 + y^2 + z^2}$.

7. (c) The direction of fastest increase is along the gradient vector. Using the result of exercise 6(c), we get the direction of fastest increase as $\nabla f(1, 1, 1) = (-2/9)(\mathbf{i} + \mathbf{j} + \mathbf{k})$.

8. The gradient vector is normal to a surface. Here, we have $f(x, y, z) = x^3y^3 + y - z + 2 = 0$, so $f_x = 3x^2y^3$, $f_y = 3x^3y^2 + 1$, and $f_z = -1$. At the point $(0, 0, 2)$, we compute $f_x = 0$, $f_y = 1$ and $f_z = -1$. Therefore, a normal vector is $\nabla f(0, 0, 2) = \mathbf{j} - \mathbf{k}$. Normalize this to get the unit normal: $(1/\sqrt{2})(\mathbf{j} - \mathbf{k})$.

13. (b) By definition, $\nabla f = (\partial f/\partial x, \partial f/\partial y, \partial f/\partial z)$, so $\nabla f = (yz\, e^{xyz}, xz\, e^{xyz}, xy\, e^{xyz})$. Given $g(t) = (6t, 3t^2, t^3)$, we differentiate each component to get $g'(t) = (6, 6t, 3t^2)$. Now, from section 2.4, we know that $(f \circ g)'(1) = \nabla f(\mathbf{g}(1)) \cdot \mathbf{g}'(1) = \exp(18t^6)(3t^5, 6t^4, 18t^3) \cdot (6, 6t, 3t^2)\,|_{t=1} = e^{18}(18 + 36 + 54) = 108e^{18}$.

17. By defintion, $\nabla f = (f_x, f_y)$. Since f is independent of y , $f_y = 0$ and given that $f(x, y) = g(x)$, we have $f_x = \partial f/\partial x = g'(x)$. Therefore, $\nabla f(x, y) = (g'(x), 0)$.

20. The direction in which the altitude is increasing most rapidly at the point (x, y) is $\nabla z(x, y) = (-2ax, -2by)$. At the point $(1, 1)$, $\nabla z(1, 1) = -2(a, b)$, so the desired direction is $-(a\mathbf{i} + b\mathbf{j})/\sqrt{a^2 + b^2}$. If a marble were released at $(1, 1)$, it will roll in the direction at which the altitude is *decreasing* most rapidly, so the marble will roll down in the direction $-\nabla z$, or $(a\mathbf{i} + b\mathbf{j})/\sqrt{a^2 + b^2}$.

24. (a) We want to maximize $f(\boldsymbol{\sigma}(t)) = (\cos t)(\sin t)$. Set the first derivative equal to 0: $df/dt = -(\sin t)(\sin t) + (\cos t)(\cos t) = 0$, so we get $\cos^2 t = \sin^2 t$ or $t = \pm(\pi/4 + n\pi)$, where $n = 0, 1, 2, \ldots$. Since $0 \le t \le 2\pi$, we only want $t = \pi/4, 3\pi/4, 5\pi/4, 7\pi/4$. Evaluating at these points, we get $f(\boldsymbol{\sigma}(\pi/4)) = f(\boldsymbol{\sigma}(5\pi/4)) = 1/2$ and $f(\boldsymbol{\sigma}(3\pi/4)) = f(\boldsymbol{\sigma}(7\pi/4)) = -1/2$. Therefore, the maximum value of f along the curve $\boldsymbol{\sigma}(t)$ is $1/2$ and the minimum value is $-1/2$.

2.6 Iterated Partial Derivatives

GOALS

1. Be able to compute iterated partial derivatives.
2. Be able to explain when mixed partials are equal.

STUDY HINTS

1. *Iterated partial derivatives.* These are higher-order derivatives, such as second and third derivatives. With several variables, higher-order derivatives may be taken with respect to different variables. The notation $\dfrac{\partial^2 f}{\partial x \partial y}$ means $\left(\dfrac{\partial}{\partial x}\right)\left(\dfrac{\partial f}{\partial y}\right)$, which is also denoted f_{yx}.

2. *Equality of mixed partials.* If the first partial derivatives are continuous, then an iterated partial derivative may be computed in any order.

3. *Warning.* Note that the theorem on equality of mixed partials requires continuous partial derivatives. If this requirement is violated, different orders of differentiation may yield different results.

4. *Evaluating partials at a given point.* Always remember to differentiate completely before substituting given values. With mixed partials, you may substitute for a variable only after you have completed differentiating in that variable.

5. *Applications.* The heat equation, the wave equation, and the potential (Laplace) equation are famous examples of how higher order derivatives occur in nature. There is normally no need to memorize these equations in a vector calculus course.

SOLUTIONS TO SELECTED EXERCISES

1. (b) The first partial derivatives are $\partial f/\partial x = e^x - 1/x^2 + e^{-y}$ and $\partial f/\partial y = -xe^{-y}$. Finding the partial derivatives of the first partial derivatives gives us these second partial derivatives:

$$\frac{\partial^2 f}{\partial x^2} = \frac{\partial}{\partial x}\left(\frac{\partial f}{\partial x}\right) = e^x + \frac{2}{x^3}\,;\quad \frac{\partial^2 f}{\partial y \partial x} = \frac{\partial}{\partial y}\left(\frac{\partial f}{\partial x}\right) = -\,e^{-y} = \frac{\partial^2 f}{\partial x \partial y}\,;$$

$$\frac{\partial^2 f}{\partial y^2} = \frac{\partial}{\partial y}\left(\frac{\partial f}{\partial y}\right) = xe^{-y}\,.$$

2. (a) If $(x, y) \neq (0, 0)$, we can compute the first partial derivatives in the usual way:

$$\frac{\partial f}{\partial x} = \frac{(y(x^2 - y^2) + 2x^2y)(x^2 + y^2) - 2x^2y(x^2 - y^2)}{(x^2 + y^2)^2} = \frac{x^4y - y^5 + 4x^2y^3}{(x^2 + y^2)^2}\,.$$

$$\frac{\partial f}{\partial y} = \frac{(x(x^2 - y^2) - 2xy^2)(x^2 + y^2) - 2xy^2(x^2 - y^2)}{(x^2 + y^2)^2} = \frac{x^5 - 4x^3y^2 - xy^4}{(x^2 + y^2)^2}\,.$$

(b) To compute the partial derivatives at $(0, 0)$, we need to use the definition of partial derivatives. First, hold y constant at 0 and differentiate with respect to x at $x_0 = 0$. Note that $f(0, 0)$ is defined to be 0. Then

$$\frac{\partial f}{\partial x}(0, 0) = \lim_{h \to 0} \frac{f(0 + h, 0) - f(0,0)}{h} = \lim_{h \to 0} \frac{\dfrac{h(0)(h^2 - 0^2)}{h^2 + 0^2} - 0}{h} = \lim_{h \to 0} \frac{0}{h}\,,$$

and by l'Hôpital's rule, $\partial f/\partial x$ at $(0, 0)$ becomes 0. Similarly, we hold x constant at 0 and differentiate with respect to y at $y_0 = 0$. Then

$$\frac{\partial f}{\partial x}(0, 0) = \lim_{h \to 0} \frac{f(0, 0 + h) - f(0, 0)}{h} = \lim_{h \to 0} \frac{\dfrac{(0)(h)(0^2 - h^2)}{0^2 + h^2} - 0}{h} = \lim_{h \to 0} \frac{0}{h} = 0\,.$$

(c) By definition, $\dfrac{\partial^2 f}{\partial x \partial y} = \dfrac{\partial}{\partial x}\left(\dfrac{\partial f}{\partial y}\right)$. First, we use $\dfrac{\partial f}{\partial y}$ from part (a) and then perform differentiation as in part (b). By definition,

$$\frac{\partial^2 f}{\partial x \partial y}(0, 0) = \lim_{h \to 0} \frac{\dfrac{\partial f}{\partial y}(0 + h, 0) - \dfrac{\partial f}{\partial y}(0, 0)}{h} = \lim_{h \to 0} \frac{\dfrac{h^5 - 4h^3(0)^2 - x(0)^4}{(h^2 + 0^2)^2} - 0}{h} =$$

$$\lim_{h \to 0} \frac{h^5/h^4}{h} = 1\,.$$

Similarly,

$$\frac{\partial^2 f}{\partial y \partial x}(0, 0) = \lim_{h \to 0} \frac{\dfrac{\partial f}{\partial x}(0, 0 + h) - \dfrac{\partial f}{\partial x}(0, 0)}{h} = \lim_{h \to 0} \frac{\dfrac{(0)^4h - h^5 + 4(0)^2h^3}{(0^2 + h^2)^2} - 0}{h} =$$

$$\lim_{h \to 0} \frac{-h^5/h^4}{h} = -1 .$$

(d) The mixed partials are not equal, which is consistent with the fact that the first partials are not continuous at $(0, 0)$.

3. (b) Rewrite the function as $z = \dfrac{2x^2 + 7x^2y}{3xy} = \dfrac{2x}{3y} + \dfrac{7x}{3}$, if $(x, y) \neq (0, 0)$.

We compute:

$$\frac{\partial z}{\partial x} = \frac{2}{3y} + \frac{7}{3} ; \quad \frac{\partial z}{\partial y} = -\frac{2x}{3y^2} ; \quad \frac{\partial^2 z}{\partial x^2} = \frac{\partial}{\partial x}\left(\frac{\partial z}{\partial x}\right) = 0 ;$$

$$\frac{\partial^2 z}{\partial x \partial y} = \frac{\partial}{\partial x}\left(\frac{\partial z}{\partial y}\right) = -\frac{2}{3y^2} = \frac{\partial}{\partial y}\left(\frac{\partial z}{\partial x}\right) = \frac{\partial^2 z}{\partial y \partial x} ; \quad \frac{\partial^2 z}{dy^2} = \frac{\partial}{\partial y}\left(\frac{\partial z}{\partial y}\right) = \frac{4x}{3y^3} .$$

At $(0, 0)$, the function is not continuous, hence it is not differentiable.

7. By definition, $\dfrac{\partial^3 f}{\partial x\, \partial y\, \partial z} = \dfrac{\partial}{\partial x}\left(\dfrac{\partial}{\partial y}\left(\dfrac{\partial f}{\partial z}\right)\right)$. Let $h = \dfrac{\partial f}{\partial z}$, so $\dfrac{\partial^3 f}{\partial x\, \partial y\, \partial z} = \dfrac{\partial}{\partial x}\left(\dfrac{\partial h}{\partial y}\right)$ and by theorem 15, this is also equal to $\dfrac{\partial}{\partial y}\left(\dfrac{\partial h}{\partial x}\right)$. Also, by theorem 15, $\dfrac{\partial h}{\partial x} = \dfrac{\partial}{\partial x}\left(\dfrac{\partial f}{\partial z}\right) = \dfrac{\partial}{\partial z}\left(\dfrac{\partial f}{\partial x}\right)$, so $\dfrac{\partial^3 f}{\partial x\, \partial y\, \partial z} = \dfrac{\partial}{\partial y}\left(\dfrac{\partial}{\partial z}\left(\dfrac{\partial f}{\partial x}\right)\right) = \dfrac{\partial^3 f}{\partial y\, \partial z\, \partial x}$.

11. (a) We are given $f(x, y) = x \arctan(x/y)$, so we compute:

$$f_x = \arctan\left(\frac{x}{y}\right) + \frac{x}{y}\,\frac{1}{1 + (x^2/y^2)} = \frac{xy}{x^2 + y^2} + \arctan\left(\frac{x}{y}\right) .$$

$$f_y = \frac{x}{1 + (x^2/y^2)} \bullet \frac{-x}{y^2} = \frac{-x^2}{x^2 + y^2} .$$

$$f_{xx} = \frac{y(x^2 + y^2) - 2x^2y}{(x^2 + y^2)^2} + \frac{1}{y} \bullet \frac{1}{1 + (x^2/y^2)} = \frac{y^3 - x^2y}{(x^2 + y^2)^2} + \frac{y}{x^2 + y^2} = \frac{2y^3}{(x^2 + y^2)^2} .$$

$$f_{xy} = f_{yx} = \frac{-2x(x^2 + y^2) + 2x^3}{(x^2 + y^2)^2} = \frac{-2xy^2}{(x^2 + y^2)^2} .$$

$$f_{yy} = \frac{x^2}{(x^2 + y^2)^2} \bullet 2y = \frac{2x^2y}{(x^2 + y^2)^2} .$$

15. We have $u_x = 3x^2 - 6xy$, so $u_{xx} = 6x - 6x = 0$. Also, $u_y = 3x^2$, so $u_{yy} = 0$. Substitution gives us $u_{xx} + u_{yy} = 0 + 0 = 0$. Thus, $u(x, y)$ satisfies Laplace's

equation and so is harmonic.

16. (b) For $u = x^2 + y^2$, we get $u_x = 2x$, so $u_{xx} = 2$. Also, $u_y = 2y$, so $u_{yy} = 2$. Substitution into Laplace's equation gives us $u_{xx} + u_{yy} = 2 + 2 \neq 0$, so $x^2 + y^2$ is *not* harmonic.
(d) For $u = y^3 + 3x^2y$, we get $u_x = 6xy$, so $u_{xx} = 6y$. Also, $u_y = 3y^2 + 3x^2$, so $u_{yy} = 6y$. Substitution into Laplace's equation gives us $u_{xx} + u_{yy} = 6y + 6y \neq 0$, so $y^3 + 3x^2y$ is *not* harmonic.

19. Given $V(x, y, z) = -GM/\sqrt{x^2 + y^2 + z^2} = -GM/r$, we compute:

$$V_x = \frac{GMx}{(x^2 + y^2 + z^2)^{3/2}}, \text{ so}$$

$$V_{xx} = GM\,\frac{(x^2 + y^2 + z^2)^{3/2} - (3/2)(x^2 + y^2 + z^2)^{1/2} \cdot 2x \cdot x}{(x^2 + y^2 + z^2)^3}$$

$$= GM\left[\frac{1}{(x^2 + y^2 + z^2)^{3/2}} - \frac{3x^2}{(x^2 + y^2 + z^2)^{5/2}}\right] = GM\left(\frac{1}{r^3} - \frac{3x^2}{r^5}\right).$$

By symmetry, $V_{yy} = GM(1/r^3 - 3y^2/r^5)$ and $V_{zz} = GM(1/r^3 - 3z^2/r^5)$. Add to get:

$$V_{xx} + V_{yy} + V_{zz} = GM\left(\frac{3}{r^3} - \frac{3(x^2 + y^2 + z^2)}{r^5}\right)$$

$$= 3GM\left(\frac{1}{r^3} - \frac{r^2}{r^5}\right) = 0.$$

2.7 Some Technical Differentiation Theorems

GOALS

1. Be able to understand the proofs of the theorems introduced in chapter 2.
2. Be able to use the triangle inequality and the definition of the limit for proving theorems.

STUDY HINTS

1. *Advanced material.* This material is usually covered in a junior or senior level math class. Potential mathematicians should read this section even if the instructor does not cover this material.

2. *How to study proofs.* As you read, you should ask yourself why each step is valid before continuing. If you feel that you have mastered the material, try to anticipate the next step.

3. *Triangle inequality.* This is used frequently in the proofs of this section. If necessary, review its statement in section 1.1.

4. *Definition of the limit.* The ε-δ definition of the limit (theorem 6) is used in most of the proofs. Note that ε is given and we need to find a δ (not the other way around!) which satisfies the definition. δ is usually found as a function of ε.

5. *New derivative definition.* In this section, an ε-δ definition of the derivative is given. It is equivalent to the definition given in section 2.3 .

SOLUTIONS TO SELECTED EXERCISES

1. Here, we have $f(x, y, z) = (e^x, \cos y, \sin z) = (f_1, f_2, f_3)$, so

$$\mathbf{D}f(x, y, z) = \begin{bmatrix} \partial f_1/\partial x & \partial f_1/\partial y & \partial f_1/\partial z \\ \partial f_2/\partial x & \partial f_2/\partial y & \partial f_2/\partial z \\ \partial f_3/\partial x & \partial f_3/\partial y & \partial f_3/\partial z \end{bmatrix} = \begin{bmatrix} e^x & 0 & 0 \\ 0 & -\sin y & 0 \\ 0 & 0 & \cos z \end{bmatrix}.$$

$\mathbf{D}f$ is a diagonal matrix when f_1 depends only on the first variable, f_2 depends only on the second, and so on. Thus, $\mathbf{D}f$ is diagonal if f_n is a function of the nth variable only.

4. A uniformly continuous function is not only continuous at all points in the domain, but more importantly, we can find *one* δ for all ε. This is different from continuity in that for continuity, we may find a δ which will work for a particular ε. A continuous function does not always have to be uniformly continuous (for examples: $f(x) = 1/x^2$ on $\mathbf{R}$ or $g(x) = 1/x$ on $(0, 1]$.)

   (a) $\mathbf{T}: \mathbf{R}^n \to \mathbf{R}^m$ is linear implies that $\| \mathbf{Tx} \| \le M \| \mathbf{x} \|$. Given $\varepsilon > 0$, $\mathbf{x}$ and $\mathbf{y} \in \mathbf{R}^n$, we have $\| \mathbf{T}(\mathbf{x} - \mathbf{y}) \| \le M \| \mathbf{x} - \mathbf{y} \|$ from exercise 2(a). Since $\mathbf{T}$ is linear, $\| \mathbf{T}(\mathbf{x} - \mathbf{y}) \| \le \| \mathbf{T}(\mathbf{x}) - \mathbf{T}(\mathbf{y}) \|$. Let $\delta = \varepsilon/M$, then for $0 \le \| \mathbf{x} - \mathbf{y} \| < \delta$, we have $\| \mathbf{T}(\mathbf{x}) - \mathbf{T}(\mathbf{y}) \| \le M\delta = \varepsilon$. Note that this proof works because δ and ε do not depend on a particular choice of a point in $\mathbf{R}^n$.

   (b) Given $\varepsilon > 0$, we want a $\delta > 0$ such that $0 < | x - x_0 | < \delta$ implies $| 1/x^2 - 1/x_0{}^2 | < \varepsilon$ for x and x_0 in $(0, 1]$. We calculate:

$$\left| \frac{1}{x^2} - \frac{1}{x_0^2} \right| = \frac{|x_0^2 - x^2|}{x^2 x_0^2} = \frac{|x - x_0|\,|x + x_0|}{x^2 x_0^2} =$$

$$|x - x_0| \cdot \left| \frac{1}{x x_0^2} + \frac{1}{x_0 x^2} \right| < |x - x_0| \frac{2}{x_0^3} < \varepsilon$$

(if x_0 is the smaller of x and x_0), i.e. ,

$$\left|\frac{1}{x^2}-\frac{1}{x_0^2}\right| \le |x-x_0| \frac{2}{\{\min(x, x_0)\}^3} .$$

Note that $\min(x, x_0) > 0$. Let $\delta < (\varepsilon/2)\{\min(x, x_0)\}^3$, then $|x - x_0| < \delta$ implies

$$\left|\frac{1}{x^2}-\frac{1}{x_0^2}\right| < \delta \cdot \frac{2}{x_0^3} < \frac{\varepsilon}{2} x_0^3 \cdot \frac{2}{x_0^3} = \varepsilon .$$

We have only shown that $f(x) = 1/x^2$ is continuous . It is not uniformly continuous since

$$\left|\frac{1}{x^2}-\frac{1}{x_0^2}\right| = \frac{|x_0^2 - x^2|}{x^2x_0^2} = \frac{|x-x_0|\,|x+x_0|}{x^2x_0^2} ,$$

so for fixed ε , any δ approaches ∞ as x_0 goes to 0 .

7. We have $f(x, y) = g(x) + h(y)$. What we want to prove is that as matrices, $\mathbf{D}f(x, y) = (\mathbf{D}g(x), \mathbf{D}h(y))$. Since g and h are differentiable at x_0 and y_0 , respectively, we have

$$\lim_{x \to x_0} \frac{|g(x) - g(x_0) - \mathbf{D}g(x - x_0)|}{|x - x_0|} = 0 \quad \text{and}$$

$$\lim_{y \to y_0} \frac{|h(y) - h(y_0) - \mathbf{D}h(y - y_0)|}{|y - y_0|} = 0 .$$

Now, use the triangle inequality:

$$|g(x) - g(x_0) + h(y) - h(y_0) - (\mathbf{D}g(x - x_0) + \mathbf{D}h(y - y_0))| \le$$

$$|g(x) - g(x_0) - \mathbf{D}g(x - x_0)| + |h(y) - h(y_0) - \mathbf{D}h(y - y_0)| .$$

Since $\|(x, y) - (x_0, y_0)\| \ge \|x - x_0\|$ and $\|(x, y) - (x_0, y_0)\| \ge \|y - y_0\|$, the sum of the two limits is greater than

$$\lim_{(x, y) \to (x_0, y_0)} \frac{|g(x) + h(y) - (g(x_0) + h(y_0)) - (\mathbf{D}g(x - x_0) + \mathbf{D}h(y - y_0))|}{\|(x, y) - (x_0, y_0)\|} ,$$

Hence this limit goes to 0 , and this satisfies the definition for differentiability of f at (x_0, y_0) .

11. Given ε , note that

$$\left|\frac{xy}{(x^2 + y^2)^{1/2}} - 0\right| \le \left|\frac{xy}{x}\right| = |y| ,$$

and we also know that $|y| \le \sqrt{x^2 + y^2}$. Let $\delta = \varepsilon$, then $\|(x, y) - (0, 0)\| = \sqrt{x^2 + y^2} < \delta$ implies that

$$\left|\frac{xy}{(x^2 + y^2)^{1/2}} - 0\right| < |y| < \varepsilon .$$

Therefore, $\lim_{(x,y) \to (0,0)} f(x, y) = 0$ and f is continuous.

16. (a) Suppose $\mathbf{y}$ is a boundary point of the open set $A \subset \mathbf{R}^n$. Then every ball centered at $\mathbf{y}$ contains points in A and points not in A. So the intersection of A and the ball $D_\varepsilon(\mathbf{y})$ is not empty. Hence

   (i) $\mathbf{y}$ is not in A since points of A have neighborhoods contained in A, and

   (ii) $\mathbf{y}$ is the limit of the sequence from A; choose $\varepsilon = 1/n$, $n = 1, 2, 3, \ldots$, then $D_{(1/n)}(\mathbf{y}) \cap A \neq \varnothing$ for $n = 1, 2, 3, \ldots$.

   Pick $\mathbf{x}_n$ to be an element of $D_{(1/n)}(\mathbf{y}) \cap A$. Then $\mathbf{x}_n$ is in A and $\| \mathbf{y} - \mathbf{x}_n \| \leq 1/n$, which goes to zero as n goes to infinity. Suppose $\mathbf{y}$ is not in A, but there is a sequence $\{\mathbf{x}_n\}$ in A converging to $\mathbf{y}$. Let $\varepsilon > 0$. Then $\mathbf{y}$ is in the intersection of $D_\varepsilon(\mathbf{y})$ and the set of all points not in A, so the right-hand side is not empty. On the other hand, there exist an N such that $n \geq N$ implies that $\| \mathbf{x}_n - \mathbf{y} \| < \varepsilon$, i.e., $n \geq N$ implies that $\mathbf{x}_n$ is in $D_\varepsilon(\mathbf{y})$. $\mathbf{x}_N$ is in A, so $\mathbf{x}_N$ is in $D_\varepsilon(\mathbf{y}) \cap A$. Hence, the right-hand side is a boundary point of A.

(b) Suppose $\lim_{\mathbf{x} \to \mathbf{y}} f(\mathbf{x}) = b$, i.e., for every $\varepsilon > 0$, there is a $\delta > 0$ such that for $\mathbf{x}$ in A, $\| \mathbf{x} - \mathbf{y} \| < \delta$ implies that $\| f(\mathbf{x}) - f(\mathbf{y}) \| < \varepsilon$. Let $\{\mathbf{x}_n\}$ be a sequence in A converging to $\mathbf{y}$. Fix $\varepsilon > 0$. Choose $\delta > 0$ as above. Choose N so that $n \geq N$ implies that $\| \mathbf{x}_n - \mathbf{y} \| < \delta$. Then $n \geq N$ implies $\| f(\mathbf{x}_n) - b \| < \varepsilon$, so the sequence $| f(\mathbf{x}_n) |$ converges to b. (We need y to be on the boundary of A to guarantee the existence of the sequence $\{\mathbf{x}_n\}$.)

   To go the other way, suppose it is not the case that $\lim_{\mathbf{x} \to \mathbf{y}} f(\mathbf{x}) = b$. Then there exist $\varepsilon > 0$ such that for every $\delta > 0$, there exist an $\mathbf{x}$ in A with $\| \mathbf{x} - \mathbf{y} \| < \delta$, but $\| f(\mathbf{x}) - b \| > \varepsilon$. Choose $\mathbf{x}_n$ in A such that $\| \mathbf{x}_n - \mathbf{y} \| < 1/n$, but $\| f(\mathbf{x}_n) - b \| > \varepsilon$ for $n = 1, 2, 3, \ldots$ (corresponding to $\delta = 1/n$). Then $\mathbf{x}_n$ converges to $\mathbf{y}$ (as in part (a)), but $f(\mathbf{x})$ does not converge to b. (Note: We have proved the "contrapositive" of the theorem. We *negate* the statement of the hypothesis (note the changes in "there exists," "for every," and the inequality signs), and prove the *negation* of the result we want to arrive at. The difference between this method and the method of proving by contradiction is that we do *not* negate the hypothesis.)

(c) Let U be open in $\mathbf{R}^m$ with $\mathbf{x}$ in U. The proof follows from part (b). Specifically,

$$\lim_{\mathbf{x}_n \to \mathbf{x}} f(\mathbf{x}_n) = f(\mathbf{x}) \text{ implies } f(\mathbf{x}_n) \to f(\mathbf{x})$$

   for any sequence $\mathbf{x}_n$ converging to x in U. By part (b), f is continuous at $\mathbf{x}$.

2.R Review Exercises for Chapter 2

SOLUTIONS TO SELECTED EXERCISES

1. (a) Since $3x^2$ and y^2 are non-negative, there is no level curve if $z < 0$. If $z = 0$, then the level curve is the origin. If $z > 0$, then the level curve is an ellipse with the major axis parallel to the y axis and the minor axis parallel to the x axis. The ellipses get larger as z increases. Put all of these level curves together to get an elliptical paraboloid.

2. (c) First, consider the surface $xyz = 0$. The surface consists of the planes $x = 0$, $y = 0$, and $z = 0$. Now, consider $xyz = 1$. When $z = k$, a positive constant, the level curve is $xy = 1/k$. So we get a hyperbola in the first and third quadrants with asymptotes on the x and y axes. The hyperbolas get closer to the origin as z gets larger. Thus, the surface in the first octant has the planes $x = 0$, $y = 0$, and $z = 0$ as asymptotes. Similarly, there is a surface in the octant where $z > 0$, $x < 0$, and $y < 0$. Now, if $z < 0$, then there is a similar surface in either of the octants where $x < 0$ and $y > 0$ or $x > 0$ and $y < 0$. The level surfaces for $xyz = c$, where c is a positive constant, consists of four similar surfaces which move further from the origin as c gets larger. If $c < 0$, the level surface is positioned in the other four octants.

3. (b) The function takes a point from $\mathbf{R}^1$ to $\mathbf{R}^2$, so the size of the derivative matrix is 2×1. The derivative is

$$\mathbf{D}f(x) = \begin{bmatrix} \partial f_1/\partial x \\ \partial f_2/\partial x \end{bmatrix} = \begin{bmatrix} 1 \\ 1 \end{bmatrix} .$$

5. We need to show the vector normal to the tangent plane of $f(x, y)$ at the point $(x_0, y_0, f(x_0, y_0))$ is *parallel* to (x_0, y_0, z_0). The partial derivatives of f are:

$$\frac{\partial f}{\partial x}(x_0, y_0) = \frac{-1}{2} \cdot 2x(1 - x^2 - y^2)^{-1/2} \Big|_{(x_0, y_0)} = \frac{-x_0}{\sqrt{1 - x_0^2 - y_0^2}} .$$

$$\frac{\partial f}{\partial y}(x_0, y_0) = \frac{-1}{2} \cdot 2y(1 - x^2 - y^2)^{-1/2} \Big|_{(x_0, y_0)} = \frac{-y_0}{\sqrt{1 - x_0^2 - y_0^2}} .$$

So the normal to the tangent plane is

$$\frac{-x_0}{\sqrt{1 - x_0^2 - y_0^2}}\mathbf{i} + \frac{-y_0}{\sqrt{1 - x_0^2 - y_0^2}}\mathbf{j} - \mathbf{k} .$$

Multiply the above through by $-(1 - {x_0}^2 - {y_0}^2)^{-1/2}$. We get $(x_0, y_0, f(x_0, y_0)) = (x_0, y_0, z_0)$. Geometrically, we are looking at the sphere: $x^2 + y^2 + z^2 = 1$. The vectors normal to the tangent planes are precisely the vectors $\mathbf{r} = (x, y, z)$. Those vectors have the direction of $\mathbf{e}_\rho$ (see exercise 7(a) in section 1.4).

7. (c) The equation of the tangent plane is $z = f(x_0, y_0) + [(\partial f/\partial x)(x_0, y_0)](x - x_0) + [(\partial f/\partial y)(x_0, y_0)](y - y_0)$. In this case, we have $f(-1, -1) = 1$; $\partial f/\partial x = y$,

$(\partial f/\partial x)(-1, -1) = -1$; $\partial f/\partial y = x$, $(\partial f/\partial y)(-1, -1) = -1$. So the equation of the tangent line is $z = 1 - 1(x + 1) - (y + 1)$ or $x + y + z = -1$.

8. (b) If $f(x, y, z) =$ constant, then the tangent plane equation is $\nabla f(\mathbf{x}_0) \cdot (\mathbf{x} - \mathbf{x}_0) = 0$, where $\mathbf{x} = (x, y, z)$. In this case, $f(x, y, z) = x^3 - 2y^3 + z^3$, so $\nabla f(\mathbf{x}) = (3x^2, -6y^2, 3z^2)$ and $\nabla f(1, 1, 1) = (3, -6, 3)$. Therefore, the tangent plane is $(3, -6, 3) \cdot (x - 1, y - 1, z - 1) = 3x - 6y + 3z = 0$ or $x - 2y + z = 0$.

11. (b) The strategy here is to find a few "paths," and compute the limit along those paths. Let $x = 2y$, then the limit as y goes to 0 is:

$$\lim_{(x,\, y) \to (0,\, 0)} \sqrt{\left|\frac{x+y}{x-y}\right|} = \lim_{y \to 0} \sqrt{\left|\frac{3y}{-y}\right|} = \sqrt{3} .$$

On the other hand, take the path $x = 4y$. Then the limit as y goes to 0 is:

$$\lim_{(x,\, y) \to (0,\, 0)} \sqrt{\left|\frac{x+y}{x-y}\right|} = \lim_{y \to 0} \sqrt{\left|\frac{5y}{3y}\right|} = \sqrt{\frac{5}{3}} \neq \sqrt{3} .$$

Since the limiting values depend upon the path taken, the limit does not exist.

12. (b) Hold y and z constant and use the chain rule to differentiate with respect to x . Then, $\partial f/\partial x = 10(x + y + z)^9$. By symmetry, $\partial f/\partial y = \partial f/\partial z = 10(x + y + z)^9$. The gradient is the vector $(\partial f/\partial x, \partial f/\partial y, \partial f/\partial z) = 10(x + y + z)^9(\mathbf{i} + \mathbf{j} + \mathbf{k})$.

16. We compute

$$\frac{\partial z}{\partial x}(1, -2) = 2x\,\Big|_{(1,\,-2)} = 2 \; ; \; \frac{\partial z}{\partial y}(1, -2) = 2y\,\Big|_{(1,\,-2)} = 4 .$$

Then, the tangent plane is $z = f(1, -2) + (\partial z/\partial x)|_{(1,\,-2)}(x - 1) + (\partial f/\partial y)|_{(1,\,-2)}(y + 2)$, or $z = 5 + 2(x - 1) - 4(y + 2)$, or $2x - 4y - z = 5$.
Geometrically, the gradient of $f(x, y) = x^2 + y^2$ at $(1, -2)$ is perpendicular to the level curve $z = 5$. The tangent plane of the graph of f is the plane that contains the line perpendicular to the gradient of f at $(1, -2)$ and lying in the horizontal plane $z = 5$, and the tangent plane has slope $\sqrt{2^2 + 4^2} = 2\sqrt{5}$ relative to the xy plane.

18. (b) The directional derivative is $\nabla f(\mathbf{x}_0) \cdot \mathbf{v}/\|\mathbf{v}\|$. We find $\nabla f(\mathbf{x}) = (y + z, x + z, y + x)$, so $\nabla f(1, 1, 2) = (3, 3, 2)$. Thus, the directional derivative is $(3, 3, 2) \cdot (10, -1, 2)/\sqrt{105} = (30 - 3 + 4)/\sqrt{105} = 31/\sqrt{105}$.

21. The bug should move in the direction of $-\nabla T(x, y)$, since this is the direction of fastest decrease. We compute $-\nabla T(x, y) = -4x\mathbf{i} + 8y\mathbf{j}$ and $-\nabla T(-1, 2) = 4\mathbf{i} + 16\mathbf{j}$, so the bug should move in the direction of $4\mathbf{i} + 16\mathbf{j}$.

25. (b) We use the chain rule to compute $g_x = \partial g/\partial x = -x(x^2 + y^2 + z^2)^{-3/2}$. Then $g_{xx} = -(x^2 + y^2 + z^2)^{-3/2} + 3x^2(x^2 + y^2 + z^2)^{-5/2}$. By symmetry, $g_{yy} = -(x^2 + y^2 + z^2)^{-3/2} + 3y^2(x^2 + y^2 + z^2)^{-5/2}$ and $g_{zz} = -(x^2 + y^2 + z^2)^{-3/2} + 3z^2(x^2 + y^2 + z^2)^{-5/2}$. Adding, we get $g_{xx} + g_{yy} + g_{zz} = -3(x^2 + y^2 + z^2)^{-3/2} +$

$$3(x^2 + y^2 + z^2)(x^2 + y^2 + z^2)^{-5/2} = 0 .$$

29. (ii)

(a) The sum rule tells us that $x^2 + y^4$ is differentiable. Also, x^2y^2 is differentiable by the product rule. Finally, $x^2y^2/(x^2 + y^4)$ is differentiable by the quotient rule since $(x, y) \neq (0, 0)$, and so $x^2 + y^4 \neq 0$. Holding y constant, we get

$$\frac{\partial f}{\partial x} = \frac{2xy^2(x^2 + y^4) - 2x(x^2y^2)}{(x^2 + y^4)^2} = \frac{2xy^6}{(x^2 + y^4)^2} .$$

Holding x constant, we get

$$\frac{\partial f}{\partial y} = \frac{2x^2y(x^2 + y^4) - 4y^3(x^2y^2)}{(x^2 + y^4)^2} = \frac{2x^4y - 2x^2y^5}{(x^2 + y^4)^2} .$$

At the origin, we must use the definition of the partial derivative:

$$\frac{\partial f}{\partial x}(0, 0) = \lim_{h \to 0} \frac{f(0 + h, 0) - f(0, 0)}{h} \lim_{h \to 0} \frac{\frac{0}{h^2} - 0}{h} = 0,$$

and similarily,

$$\frac{\partial f}{\partial y}(0, 0) = \lim_{h \to 0} \frac{f(0, 0 + h) - f(0, 0)}{h} = \lim_{h \to 0} \frac{\frac{0}{h^4}}{h} = 0.$$

(b) The partials exist at $(0, 0)$ and

$$\lim_{(x, y) \to (0, 0)} \frac{f(x, y)}{\| (x, y) - (0, 0) \|} = 0,$$

so by the definition of differentiability (equation (2), section 2.3), f is differentiable at $(0, 0)$. F is differentiable at all other points because the partials are continuous. Thus f is differentiable. However, as (x, y) approaches the origin, $\partial f/\partial x$ and $\partial f/\partial y$ do not approach 0 (take, for example, the path $x = y$) thus the partial derivatives are not continuous.

30. (b) The gradient vector is $\nabla f = (\partial f/\partial x, \partial f/\partial y)$. If $(x, y) \neq (0, 0)$, then

$$\frac{\partial f}{\partial x} = y \sin\left(\frac{1}{x^2 + y^2}\right) - \frac{2x^2y}{(x^2 + y^2)^2} \cos\left(\frac{1}{x^2 + y^2}\right)$$

and by symmetry,

$$\frac{\partial f}{\partial y} = x \sin\left(\frac{1}{x^2 + y^2}\right) - \frac{2xy^2}{(x^2 + y^2)^2} \cos\left(\frac{1}{x^2 + y^2}\right) .$$

Now, if $(x, y) = (0, 0)$, then the definition of the partial derivative gives us

$$\frac{\partial f}{\partial x}(0, 0) = \lim_{h \to 0} \frac{f(0 + h, 0) - f(0, 0)}{h} = \lim_{h \to 0} \frac{(0 + h)(0) \sin\left(\frac{1}{(0 + h)^2 + (0)^2}\right)}{h} =$$

$$\lim_{h \to 0} \frac{0}{h} = 0 .$$

Similarly, $(\partial f/\partial y)(0, 0) = 0$. Therefore, if $(x, y) \neq (0, 0)$, then

$$\nabla f(x, y) = \left[y \sin\left(\frac{1}{x^2 + y^2}\right) - \frac{2x^2y}{(x^2 + y^2)^2} \cos\left(\frac{1}{x^2 + y^2}\right) \right] \mathbf{i} +$$

$$\left[x \sin\left(\frac{1}{x^2 + y^2}\right) - \frac{2xy^2}{(x^2 + y^2)^2} \cos\left(\frac{1}{x^2 + y^2}\right) \right] \mathbf{j},$$

and $\nabla f(0, 0) = 0\mathbf{i} + 0\mathbf{j}$.

33. (b) The directional derivative is $\nabla f(\mathbf{x}_0) \cdot \mathbf{v}$. Here,

$$\frac{\partial f}{\partial x} = -\frac{x}{\sqrt{x^2 + y^2}} \sin\left(\sqrt{x^2 + y^2}\right)$$

and by symmetry,

$$\frac{\partial f}{\partial y} = -\frac{y}{\sqrt{x^2 + y^2}} \sin\left(\sqrt{x^2 + y^2}\right).$$

Therefore, $\nabla f(1, 1) = ((-1/\sqrt{2})\sin(\sqrt{2}), (-1/\sqrt{2})\sin(\sqrt{2}))$ and the directional derivative is

$$((-1/\sqrt{2})\sin(\sqrt{2}), (-1/\sqrt{2})\sin(\sqrt{2})) \cdot (1/\sqrt{2}, 1/\sqrt{2}) = -\sin(\sqrt{2}).$$

37. (b) Directly, we first compute $g(u) = f(\mathbf{h}(u)) = \sin^2 3u + \cos 8u$. Then $dg/du = 6(\sin 3u)(\cos 3u) - 8 \sin 8u$. When $u = 0$, $dg/du = 0$. By the chain rule,

$$\frac{dg}{du} = \mathbf{D}f(x, y) \cdot \mathbf{D}\mathbf{h}(u) = [2x, 1] \begin{bmatrix} 3 \cos 3u \\ -8 \sin 8u \end{bmatrix} = [2 \sin 3u, 1] \begin{bmatrix} 3 \cos 3u \\ -8 \sin 8u \end{bmatrix}$$
$$= 6(\sin 3u)(\cos 3u) - 8 \sin 8u.$$

Again, when $u = 0$, $dg/du = 0$.

39. The normal to the surface $f(x, y, z) = x^2 + 2y^2 + 3z^2 = 6$ is $(\partial f/\partial x, \partial f/\partial y, \partial f/\partial z) = (2x, 4y, 6z) = (x, 2y, 3z)$. At $(1, 1, 1)$, the normal is $(1, 2, 3)$, so the unit normal is $(1, 2, 3)/\sqrt{14}$, the direction of flight. The velocity of the particle is the speed times the direction, or $10(1, 2, 3)/\sqrt{14}$. The position of the particle at any time can be found by finding the equation of the line through $(1, 1, 1)$ with the direction $10(1, 2, 3)/\sqrt{14}$, and it is $(1, 1, 1) + 10t\,(1, 2, 3)/\sqrt{14}$. This implies that $x = 1 + 10t/\sqrt{14}$, $y = 1 + 20t/\sqrt{14}$, and $z = 1 + 30t/\sqrt{14}$. At some time T, the particle is on the sphere $x^2 + y^2 + z^2 = 103$, which means that $(1 + 10T/\sqrt{14})^2 + (1 + 20T/\sqrt{14})^2 + (1+30T/\sqrt{14})^2 = 103$. Simplifying, we get $3 + (120/\sqrt{14})T + 100T^2 - 103 = 0$. Solving for T using the quadratic formula, and taking the positive T only, we get $T = (-3 + \sqrt{359})/5\sqrt{14}$.

42. (a) By substitution, $z = (x + y)(x - y) = x^2 - y^2$, so $\partial z/\partial x = 2x$ and $\partial z/\partial y = -2y$.

(b) By the chain rule, we have

$$\frac{\partial z}{\partial x} = \frac{\partial z}{\partial u} \cdot \frac{\partial u}{\partial x} + \frac{\partial z}{\partial v} \cdot \frac{\partial v}{\partial x} = v(1) + u(1) = v + u = 2x.$$

Also, we have

$$\frac{\partial z}{\partial y} = \frac{\partial z}{\partial u} \bullet \frac{\partial u}{\partial y} + \frac{\partial z}{\partial v} \bullet \frac{\partial v}{\partial y} = v(1) + u(-1) = v - u = -2y\ .$$

45. Let $u(t) = f(t)g(t)$ and $h(u) = e^u$. By the chain rule, we have

$$\frac{dh}{dt} = \frac{dh}{du} \bullet \frac{du}{dt} = e^u\left[\frac{df}{dt}\, g(t) + f(t)\,\frac{dg}{dt}\right] = \left[\frac{df}{dt}\, g(t) + f(t)\,\frac{dg}{dt}\right] \exp[f(t)g(t)]\ .$$

50. The velocity is defined as the derivative of displacement. Therefore, we want to compute $\partial u/\partial t = -6\cos(x - 6t) + 6\cos(x + 6t)$. When $t = 1/2$, $x = 1$, and the velocity is $\partial u/\partial t = -6\cos(-2) + 6\cos(4)$. Since $\cos(-x) = \cos x$, the velocity is $6(\cos(2) + \cos(4))$.

53. (a) As given, P is a function of T and V. We can also write

$$T = \left(P + \frac{\alpha}{V^2}\right)\left(\frac{V - \beta}{R}\right),$$

so T is a function of P and V. Finally, we can write

$$P = \frac{RTV^2 - \alpha(V - \beta)}{V^2(V - \beta)}\ .$$

Upon rearranging, we get

$$PV^3 - (P\beta - RT)V^2 + \alpha V - \alpha\beta = 0\ .$$

Since this is a cubic equation in V, it is theoretically possible to write V in terms of P and T. Therefore, any two of V, P, or T determines the third variable.

(b) From the equation for T, we hold V constant and get

$$(\partial T/\partial P) = (V - \beta)/R\ .$$

From the equation for P, we hold T constant and get

$$(\partial P/\partial V) = -RT/(V - \beta)^2 + 2V\alpha/V^4 = -RT/(V - \beta)^2 + 2\alpha/V^3\ .$$

Now, hold P constant and differentiate the equation for P by implicit differentiation. We get

$$0 = \frac{R(V - \beta) - \frac{\partial V}{\partial T}(RT)}{(V - \beta)^2} + \frac{2V\,\frac{\partial V}{\partial T}\,\alpha}{V^4} = \frac{R}{V - \beta} - \frac{\partial V}{\partial T}\left[\frac{RT}{(V - \beta)^2} - \frac{2\alpha}{V^3}\right],$$

or equivalently,

$$\frac{\partial V}{\partial T} = \frac{R}{(V - \beta)\left(\frac{RT}{(V - \beta)^2} - \frac{2\alpha}{V^3}\right)}\ .$$

(c) Using the results of part (b), we get

$$\left(\frac{\partial T}{\partial P}\right)\left(\frac{\partial P}{\partial V}\right)\left(\frac{\partial V}{\partial T}\right) = \left(\frac{V - \beta}{R}\right)\left(\frac{-RT}{(V - \beta)^2} + \frac{2\alpha}{V^3}\right)\left(\frac{R}{(V - \beta)\left(\frac{RT}{(V - \beta)^2} - \frac{2\alpha}{V^3}\right)}\right)$$

$$= -1\ .$$

54. (a) The question is asking for the directional derivative in the direction of a *unit* vector. Here, our unit vector is $(1, 1)/\sqrt{2}$. Also, $\nabla h(x, y) = (-0.00130x,$

$-0.00048y)$ and $\nabla h(-2, -4) = (0.00260, 0.00196)$. Therefore, the directional derivative at $(-2, -4)$ in the direction of $(1, 1)/\sqrt{2}$ is $\nabla h(-2, -4) \cdot (1, 1)/\sqrt{2}$ $= 0.00456/\sqrt{2}$. This means that the height increases $(0.00456/\sqrt{2}\,)$ miles per horizontal mile traveled.

(b) The direction of the steepest upward path is $\nabla h(-2, -4) = (0.00260, 0.00196)$.

CHAPTER 3
VECTOR-VALUED FUNCTIONS

3.1 Paths and Velocity

GOALS

1. Given a path, be able to compute the velocity vector, the acceleration vector, and the speed at a given point.

2. Be able to find a tangent line for a given path.

STUDY HINTS

1. *Paths.* A path is a "formula" which describes a curve in space. The picture of the path, which we can draw on paper, is called the image of the path or the curve of the path.

2. *Path images.* Often, it is convenient to express a path in terms of x and y when you want to know the image of a path. This is done by eliminating the parameter. For example, $(x, y) = (t^2, t^4)$ means $t = \sqrt{x}$, so $y = t^4 = (\sqrt{x})^4 = x^2$. Caution: in this example, $x = t^2$, so x is always nonnegative.

3. *Circular functions.* If a path is parametrized by some form of $\cos t$ and $\sin t$, the parameter can usually be eliminated by squaring and adding. Use the identity $\cos^2 t + \sin^2 t = 1$ and other trigonometric identities.

4. *Velocity, acceleration, and speed.* The velocity vector's components are the first derivatives of the components of the path. Speed is the length of the velocity vector. The second derivatives make up the acceleration. Beware that the derivative of speed is *not* the acceleration. Note that speed is a scalar, while velocity and acceleration are vectors.

5. *Tangent lines.* It is easy to find a tangent line if you remember that a line can be described as $\mathbf{a} + t\mathbf{v}$. The vector $\mathbf{a}$ is chosen to be a point on the path and $\mathbf{v}$ is the velocity vector $\boldsymbol{\sigma}'(t)$.

6. *Alternative solution to example 6.* A point on the path *and* the tangent line is $\boldsymbol{\sigma}(\pi) = (-1, 0, \pi)$. From time π to 2π, the particle moves along the tangent vector, so the displacement is $(2\pi - \pi)(\mathbf{v}(t))$, where $\mathbf{v}(t)$ is the tangent vector. Thus, the answer is $(-1, 0, \pi) + \pi(0, -1, 1)$.

7. *Newton's second law.* This states that $\mathbf{F} = m\mathbf{a}$, where $\mathbf{F}$ is force and $\mathbf{a}$ is the acceleration. You should remember this.

8. *Example 7*. This example derives three famous equations from astronomy. These formulas probably don't need to be memorized. Ask your instructor.

SOLUTIONS TO SELECTED EXERCISES

1. (b) Take the first derivative of each component to get $\boldsymbol{\sigma}'(t) = (e^t, -\sin t, \cos t)$. Evaluate at $t = 0$ to get $\boldsymbol{\sigma}'(0) = (1, 0, 1)$.

2. (b) We have $\boldsymbol{\sigma}(t) = (\sin 3t, \cos 3t, 2t^{3/2})$, so $\boldsymbol{\sigma}(1) = (\sin 3, \cos 3, 2)$. Taking the first derivatives, we get the velocity vector $\mathbf{v}(t) = \boldsymbol{\sigma}'(t) = (3 \cos 3t, -3 \sin 3t, 3\sqrt{t})$, so $\boldsymbol{\sigma}'(1) = (3 \cos 3, -3 \sin 3, 3)$. The second derivatives give us the acceleration vector $\mathbf{a}(t) = \boldsymbol{\sigma}''(t) = (-9 \sin 3t, -9 \cos 3t, (3/2)t^{-1/2})$, so $\boldsymbol{\sigma}''(1) = (-9 \sin 3, -9 \cos 3, 3/2)$. The equation of the tangent line is $\mathbf{l}(t) = \boldsymbol{\sigma}(1) + t\boldsymbol{\sigma}'(1)$ or $\mathbf{l}(t) = (\sin 3, \cos 3, 2) + t(3 \cos 3, -3 \sin 3, 3)$.

3. (b) Take the first derivative of each component to get the velocity vector $\boldsymbol{\sigma}'(t) = \mathbf{v}(t) = (t \cos t + \sin t, -t \sin t + \cos t, \sqrt{3})$. The acceleration vector is composed of the second derivatives, so $\boldsymbol{\sigma}''(t) = \mathbf{a}(t) = (-t \sin t + 2 \cos t, -t \cos t - 2 \sin t, 0)$. Therefore, $\mathbf{v}(0) = (0, 1, \sqrt{3})$, $\mathbf{a}(0) = (2, 0, 0)$, and the tangent line is $\mathbf{l}(\lambda) = \boldsymbol{\sigma}(0) + \lambda\mathbf{v}(0) = \mathbf{0} + \lambda(0, 1, \sqrt{3}) = \lambda(0, 1, \sqrt{3})$.

5. We want $\mathbf{F}(0) = m\mathbf{a}(0)$, where $m = 1$. The acceleration vector is $\mathbf{a}(t) = \mathbf{r}''(t) = (-\cos t, -4 \sin 2t)$, so $\mathbf{a}(0) = (-1, 0)$. Thus, $\mathbf{F}(0) = -\mathbf{i}$ g-cm/s^2 $= -0.001\mathbf{i}$ newton.

8. (b)

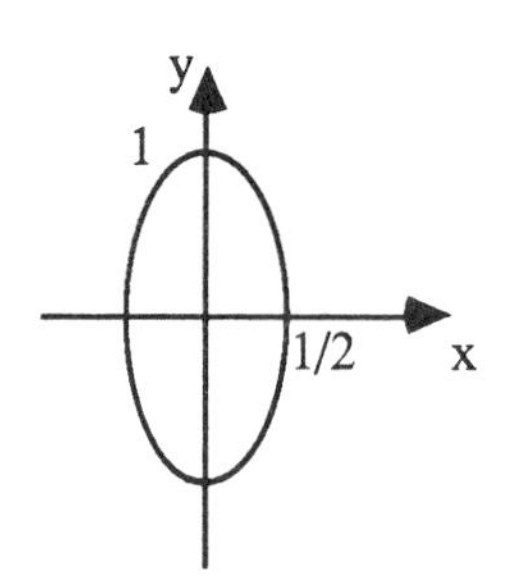

Let $X = 2x$, so $X^2 + y^2 = 1$. Recognizing this as a circle, we let $X = \cos t$ and $y = \sin t$, so $x = X/2 = (\cos t)/2$. Our path $\boldsymbol{\sigma}(t)$ is described by $(x, y) = ((\cos t)/2, \sin t)$.

12. (b) Let $\boldsymbol{\sigma}(t) = \sigma_1(t)\mathbf{i} + \sigma_2(t)\mathbf{j} + \sigma_3(t)\mathbf{k}$ and $\boldsymbol{\rho}(t) = \rho_1(t)\mathbf{i} + \rho_2(t)\mathbf{j} + \rho_3(t)\mathbf{k}$, so

$$\sigma(t) \times \rho(t) = \begin{vmatrix} \mathbf{i} & \mathbf{j} & \mathbf{k} \\ \sigma_1(t) & \sigma_2(t) & \sigma_3(t) \\ \rho_1(t) & \rho_2(t) & \rho_3(t) \end{vmatrix}$$

$$= [\sigma_2(t)\,\rho_3(t) - \sigma_3(t)\,\rho_2(t)]\mathbf{i} + [\sigma_3(t)\,\rho_1(t) - \sigma_1(t)\,\rho_3(t)]\mathbf{j} +$$
$$= [\sigma_1(t)\,\rho_2(t) - \sigma_2(t)\rho_1(t)]\mathbf{k}\,.$$

By the product rule, the left-hand side is

$$(d/dt)\,[\boldsymbol{\sigma}(t) \times \boldsymbol{\rho}(t)] = [\sigma_2'(t)\,\rho_3(t) + \sigma_2(t)\,\rho_3'(t) - \sigma_3'(t)\,\rho_2(t) - \sigma_3(t)\,\rho_2'(t)]\mathbf{i} +$$
$$[\sigma_3'(t)\,\rho_1(t) + \sigma_3(t)\,\rho_1'(t) - \sigma_1'(t)\,\rho_3(t) - \sigma_1(t)\,\rho_3'(t)]\mathbf{j} +$$
$$[\sigma_1'(t)\,\rho_2(t) + \sigma_1(t)\,\rho_2'(t) - \sigma_2'(t)\,\rho_1(t) - \sigma_2(t)\,\rho_1'(t)]\mathbf{k}\,.$$

Also, we compute

$$\left(\frac{d\boldsymbol{\sigma}}{dt}\right) \times \boldsymbol{\rho}(t) = \begin{vmatrix} \mathbf{i} & \mathbf{j} & \mathbf{k} \\ \sigma_1'(t) & \sigma_2'(t) & \sigma_3'(t) \\ \rho_1(t) & \rho_2(t) & \rho_3(t) \end{vmatrix}$$
$$= [\sigma_2'(t)\rho_3(t) - \sigma_3'(t)\rho_2(t)]\mathbf{i} +$$
$$[\sigma_3'(t)\,\rho_1(t) - \sigma_1'(t)\,\rho_3(t)]\mathbf{j} + [\sigma_1'(t)\,\rho_2(t) - \sigma_2'(t)\,\rho_1(t)]\mathbf{k}\,.$$

Similarly,

$$\boldsymbol{\sigma}(t) \times (d\boldsymbol{\rho}/dt) = [\sigma_2(t)\,\rho_3'(t) - \sigma_3(t)\,\rho_2'(t)]\mathbf{i} +$$
$$[\sigma_3(t)\,\rho_1'(t) - \sigma_1(t)\,\rho_3'(t)]\mathbf{j} + [\sigma_1(t)\,\rho_2'(t) - \sigma_2(t)\,\rho_1'(t)]\mathbf{k}\,.$$

Adding and comparing shows that

$$(d/dt)\,[\boldsymbol{\sigma}(t) \times \boldsymbol{\rho}(t)\,] = (d\boldsymbol{\sigma}/dt) \times \boldsymbol{\rho}(t) + \boldsymbol{\sigma}(t) \times (d\boldsymbol{\rho}/dt)\,.$$

14. Using exercise 12(b), we get

$$(d/dt)[m\,\boldsymbol{\sigma}(t) \times \mathbf{v}(t)] = m(d/dt)[\boldsymbol{\sigma}(t) \times \mathbf{v}(t)] = m[d\boldsymbol{\sigma}/dt \times \mathbf{v}(t) + \boldsymbol{\sigma}(t) \times d\mathbf{v}/dt]\,.$$

But $d\boldsymbol{\sigma}/dt = \mathbf{v}(t)$, $d\mathbf{v}/dt = \mathbf{a}(t)$, and $\mathbf{v}(t) \times \mathbf{v}(t) = \mathbf{0}$, so the expression reduces to

$$m[\mathbf{0} + \boldsymbol{\sigma}(t) \times \mathbf{a}(t)]\,.$$

If c is a constant,

$$c(\mathbf{u} \times \mathbf{w}) = c\mathbf{u} \times \mathbf{w} = \mathbf{u} \times c\mathbf{w}\,,$$

so

$$m[\boldsymbol{\sigma}(t) \times \mathbf{a}(t)] = \boldsymbol{\sigma}(t) \times m\mathbf{a}(t) = \boldsymbol{\sigma}(t) \times \mathbf{F}(\boldsymbol{\sigma}(t))\,.$$

If $\mathbf{F}(\boldsymbol{\sigma}(t))$ and $\boldsymbol{\sigma}(t)$ are parallel, their cross product is $\mathbf{0}$, so the angular momentum is constant.
This is the case of planetary motion, since

$$\mathbf{F}\,(\sigma\,(t)) = \frac{mM\boldsymbol{\sigma}(t)}{\|\boldsymbol{\sigma}(t)\|^3}\,,$$

we see that $\mathbf{F}(\boldsymbol{\sigma}(t))$ is a multiple of $\boldsymbol{\sigma}(t)$, so it is parallel to $\boldsymbol{\sigma}(t)$.

3.2 Arc Length

GOALS

1. Be able to compute the arc length of a given segment of a path.

STUDY HINTS

1. *Notation.* Often **s** is used to denote a path in space, rather than $\boldsymbol{\sigma}$.

2. *Arc length.* This is just the length of a curve. Lengths may be added together, so we may compute the arc length of curves which are not differentiable at finitely many points.

3. *Arc length formula.* You may find it easier to remember that arc length is the integral of speed (*not* velocity!). This makes sense because arc length gives the distance traveled. In any case, you should know that the formula is

$$\ell(\boldsymbol{\sigma}) = \int_a^b \|\boldsymbol{\sigma}'(t)\| \, dt .$$

4. *Integration tricks.* (a) Due to the nature of the arc length formula, follow the advise of the footnote on page 208 and memorize the formula

$$\int \sqrt{x^2 + a^2}\, dx = \left(\frac{1}{2}\right)\left[\, x\sqrt{x^2 + a^2} + a^2 \log\left(x + \sqrt{x^2 + a^2}\right)\right] + C .$$

This formula may be derived by trigonometric substitution if you do not wish to memorize it. Ask your instructor if this and similar formulas will be provided on an exam.
(b) Look for perfect squares. Radicals can be eliminated from the integrand if a perfect squares occurs.

5. *Positive lengths only.* If you compute a negative or a zero arc length, then you made a mistake. Arc length is always positive.

6. *Riemann sum derivation.* If the derivation does not make sense now, you should return to this discussion after Riemann sums are explained more thoroughly in chapter 5. Understanding the derivation of formulas gives much more insight into the theory.

7. *Parametrization warning.* If you need to parametrize a curve, be sure the curve is traversed once and only once. The orientation is also important, and this will become especially significant in chapter 7. The bad consequences of an incorrectparametrization are illustrated in example 1.

SOLUTIONS TO SELECTED EXERCISES

1. (b) From exercise 2(b), section 3.1, $\boldsymbol{\sigma}'(t) = (3 \cos 3t, -3 \cos 3t, 3\sqrt{t}\ )$, so

$\| \boldsymbol{\sigma}'(t) \| = [(3 \cos 3t)^2 + (-3 \sin 3t)^2 + (\sqrt{t})^2]^{1/2} = \sqrt{9 + 9t} = 3\sqrt{1 + t}$. So the arc length is

$$\int_0^1 3\sqrt{1+t}\, dt = 3 \cdot \frac{2}{3} \cdot (1+t)^{3/2} \Big|_0^1 = 2(2^{3/2} - 1) = 4\sqrt{2} - 2 .$$

(g) Given $\boldsymbol{\sigma}(t) = (t, t, (2/3)t^{3/2})$, we compute $\boldsymbol{\sigma}'(t) = (1, 1, \sqrt{t})$;
$\| \boldsymbol{\sigma}'(t) \| = \sqrt{1 + 1 + t} = \sqrt{2 + t}$, so the arc length is

$$\int_{t_0}^{t_1} \sqrt{2+t}\, dt = \frac{2}{3}(2+t)^{3/2} \Big|_{t_0}^{t_1} = \frac{2}{3}[(2+t_1)^{3/2} - (2+t_0)^{3/2}] .$$

4. Given $\boldsymbol{\sigma}(t) = (t, t \sin t, t \cos t)$, we compute $\boldsymbol{\sigma}'(t) = (1, \sin t + t \cos t, \cos t - t \sin t)$. Then $\| \boldsymbol{\sigma}'(t) \| = [1 + (\sin t + t \cos t)^2 + (\cos t - t \sin t)^2]^{1/2} = \sqrt{2 + t^2}$. Since $\boldsymbol{\sigma}(0) = (0, 0, 0)$ and $\boldsymbol{\sigma}(\pi) = (\pi, 0, -\pi)$, we want the arc length on $[0, \pi]$, so we need to compute:

$$\int_0^{\pi} \sqrt{2+t^2}\, dt .$$

This may be integrated by the method of trigonometric substitution: one would let $\tan\theta = \sqrt{2}\, t$, and then perform an integration involving $\sec^3\theta$. This is left as an exercise. An alternative is to use the formula given in the integration tables:

$$\int_0^{\pi} \sqrt{2+t^2}\, dt = \frac{1}{2}\left[t\sqrt{t^2+2} + 2\ln\left(t + \sqrt{t^2+2}\right)\right] \Big|_0^{\pi}$$

$$= \frac{1}{2}\left[\pi\sqrt{\pi^2+2} + 2\ln\left(\pi + \sqrt{\pi^2+2}\right) - 2\ln\sqrt{2}\right] .$$

6. (a) Since $\mathbf{T}(t) \cdot \mathbf{T}(t) = \| \mathbf{T}(t) \|^2 = 1$, we can differentiate both sides of $\mathbf{T}(t) \cdot \mathbf{T}(t) = 1$. By the product rule for dot products (see exercise 12(a), section 3.1), $(d/dt)(\mathbf{T}(t) \cdot \mathbf{T}(t)) = (d/dt)(1) = 2\, \mathbf{T}(t) \cdot \mathbf{T}'(t) = 0$, which implies that $\mathbf{T}(t) \cdot \mathbf{T}'(t) = 0$.

(b) Beginning with $\mathbf{T}(t) = \boldsymbol{\sigma}'(t) / \| \boldsymbol{\sigma}'(t) \|$, we differentiate with respect to t , using the quotient rule:

$$\mathbf{T}'(t) = \frac{d}{dt}\left(\frac{\boldsymbol{\sigma}'(t)}{\| \boldsymbol{\sigma}'(t) \|}\right) = \frac{\boldsymbol{\sigma}''(t) \| \boldsymbol{\sigma}'(t) \| - (d/dt) \| \boldsymbol{\sigma}'(t) \| \boldsymbol{\sigma}'(t)}{\| \boldsymbol{\sigma}'(t) \|^2} .$$

Recall that $\| \boldsymbol{\sigma}'(t) \|^2 = \boldsymbol{\sigma}'(t) \cdot \boldsymbol{\sigma}'(t)$, so

$$\frac{d}{dt} \| \boldsymbol{\sigma}'(t) \| = \frac{d}{dt} \sqrt{\boldsymbol{\sigma}'(t) \cdot \boldsymbol{\sigma}'(t)} = \frac{1}{2} (\boldsymbol{\sigma}'(t) \cdot \boldsymbol{\sigma}'(t))^{-1/2} (2\boldsymbol{\sigma}'(t) \cdot \boldsymbol{\sigma}''(t))$$

$$= \frac{\boldsymbol{\sigma}'(t) \cdot \boldsymbol{\sigma}''(t)}{\| \boldsymbol{\sigma}'(t) \|} .$$

Substitution yields

$$\mathbf{T}'(t) = \frac{\boldsymbol{\sigma}''(t)}{\|\boldsymbol{\sigma}'(t)\|^2}\|\boldsymbol{\sigma}'(t)\| - \frac{\boldsymbol{\sigma}'(t)\cdot\boldsymbol{\sigma}''(t)}{\|\boldsymbol{\sigma}'(t)\|^3}\boldsymbol{\sigma}'(t)$$

$$= \frac{\boldsymbol{\sigma}''(t)\|\boldsymbol{\sigma}'(t)\|^2 - (\boldsymbol{\sigma}'(t)\cdot\boldsymbol{\sigma}''(t))\boldsymbol{\sigma}'(t)}{\|\boldsymbol{\sigma}'(t)\|^3}.$$

11. (a) From the definitions of k and **N** in exercises 7(b) and 8,

$$\frac{d\mathbf{T}}{ds} = \mathbf{T}'(s) = \|\mathbf{T}'(s)\|\frac{\mathbf{T}'(s)}{\|\mathbf{T}'(s)\|} = k\mathbf{N}.$$

The vectors **T**, **N**, and **B** have unit length and form a right-handed system of mutually orthogonal vectors, so we have

$$\mathbf{T}\times\mathbf{N} = \mathbf{B},\ \mathbf{N}\times\mathbf{B} = \mathbf{T},\ \text{and}\ \mathbf{B}\times\mathbf{T} = \mathbf{N}.$$

Differentiate, using the product rule for cross products, to get:

$$\frac{d\mathbf{N}}{ds} = \frac{d}{ds}(\mathbf{B}\times\mathbf{T}) = \frac{d\mathbf{B}}{ds}\times\mathbf{T} + \mathbf{B}\times\frac{d\mathbf{T}}{ds} = \frac{d\mathbf{B}}{ds}\times\mathbf{T} - \frac{d\mathbf{T}}{ds}\times\mathbf{B}.$$

Using the fact that $d\mathbf{B}/ds = -\tau\mathbf{N}$ (exercise 9) along with the results derived earlier in this exercise and then factoring out a constant from the cross products, we get

$$-\tau(\mathbf{N}\times\mathbf{T}) - k(\mathbf{N}\times\mathbf{B}) = -\tau(-\mathbf{B}) - k\mathbf{T} = -k\mathbf{T} + \tau\mathbf{B}.$$

Finally,

$$\frac{d\mathbf{B}}{ds} = \frac{d}{ds}(\mathbf{T}\times\mathbf{N}) = \frac{d\mathbf{T}}{ds}\times\mathbf{N} + \mathbf{T}\times\frac{d\mathbf{N}}{ds} = k\mathbf{N}\times\mathbf{N} + \mathbf{T}\times(-k\mathbf{T} + \tau\mathbf{B})$$

$$= k\mathbf{N}\times\mathbf{N} - k\mathbf{T}\times\mathbf{T} + \tau\mathbf{T}\times\mathbf{B} = -\tau\mathbf{N}$$

since $\mathbf{N}\times\mathbf{N} = \mathbf{T}\times\mathbf{T} = \mathbf{0}$.

(b) First, we let $\boldsymbol{\omega} = (\omega_1, \omega_2, \omega_3)$ and then compute

$$\boldsymbol{\omega}\times\begin{pmatrix}\mathbf{T}\\ \mathbf{N}\\ \mathbf{B}\end{pmatrix} = \begin{vmatrix}\mathbf{i} & \mathbf{j} & \mathbf{k}\\ \omega_1 & \omega_2 & \omega_3\\ \mathbf{T} & \mathbf{N} & \mathbf{B}\end{vmatrix}$$

$$= (\mathbf{B}\omega_2 - \mathbf{N}\omega_3)\mathbf{i} - (\mathbf{B}\omega_1 - \mathbf{T}\omega_3)\mathbf{j} + (\mathbf{N}\omega_1 - \mathbf{T}\omega_2)\mathbf{k}$$

$$= \frac{d\mathbf{T}}{ds}\mathbf{i} + \frac{d\mathbf{N}}{ds}\mathbf{j} + \frac{d\mathbf{B}}{ds}\mathbf{k}.$$

Equating the **i** components, we have $d\mathbf{T}/ds = k\mathbf{N} = \mathbf{B}\omega_2 - \mathbf{N}\omega_3$, so $\omega_2 = 0$ and $\omega_3 = -k$. Similarly, we get $\omega_1 = -\tau$, so

$$\boldsymbol{\omega} = (-\tau, 0, -k).$$

3.3 Vector Fields

GOALS

1. Be able to sketch simple vector fields.

2. Understand the relationship between flow lines and a vector field.

STUDY HINTS

1. *Vector field.* This is a mapping from $\mathbf{R}^n$ to $\mathbf{R}^n$. Note that the dimensions of the spaces are equal. Each point $\mathbf{x}$ in the domain is assigned a vector. To depict a vector field, we draw the assigned vector originating from the point $\mathbf{x}$.

2. *Scalar field.* This differs from a vector field in that each point $\mathbf{x}$ in the domain is assigned a scalar, *not* a vector. An example is the annual rainfall at each point on the earth's surface. The wind velocity at any instant of time is an example of a vector field.

3. *Gradient vs. vector field.* All gradient fields are vector fields, but not all vector fields are gradient fields. For example, the vector field $\mathbf{i} + x\mathbf{j}$ is not a gradient field because you cannot find an f such that $\partial f/\partial x = 1$ and $\partial f/\partial y = x$. If a vector field $P\mathbf{i} + Q\mathbf{j}$ is a gradient field then $\partial P/\partial y = \partial Q/\partial x$. This statement comes from the fact that $\partial^2 f/\partial x \partial y = \partial^2 f/\partial y \partial x$ for a well-behaved function f.

4. *Flow lines.* This is the path a particle would take if it was free to move along the vectors in the field. Thinking of the vector field as velocity, the flow lines would show displacement. A formula description of a flow line can be obtained by integrating each component of a vector field (or solving a system of differential equations).

SOLUTIONS TO SELECTED EXERCISES

2. (b) At each point (x, y), draw a little arrow in the direction of $(x, -y)$, then connect the little arrows to get flowlines. Alternatively, one can solve the system of differential equations $da/dt = x$, $dy/dt = -y$ and write y in term of x, then plot the flowlines $y = C/x$ for various constants C. Our computed generated sketch is shown on the next page.

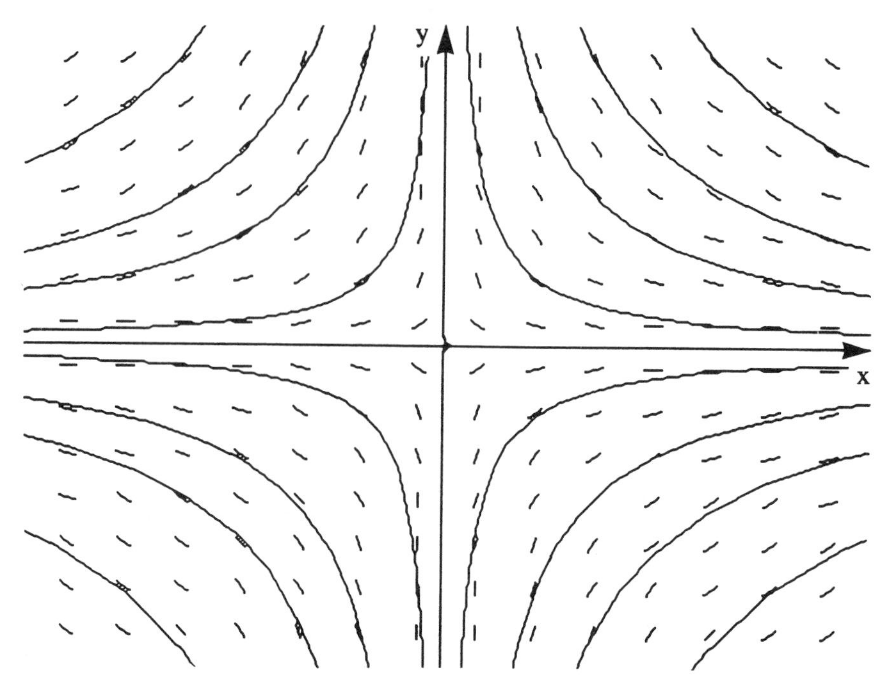

4. First, sketch the level surface $V(x, y) = (x + y)/(x^2 + y^2) = 1$. Rearranging and completing the square, we get $(x^2 - x + 1/4) + (y^2 - y + 1/4) = (x - 1/2)^2 + (y - 1/2)^2 = 1/2$. Therefore, the level surface $V(x, y) = 1$ is a circle of radius $\sqrt{1/2}$, centered at $(1/2, 1/2)$. The graph of $-\nabla V$ is two sets of circles passing through the origin, as shown below.

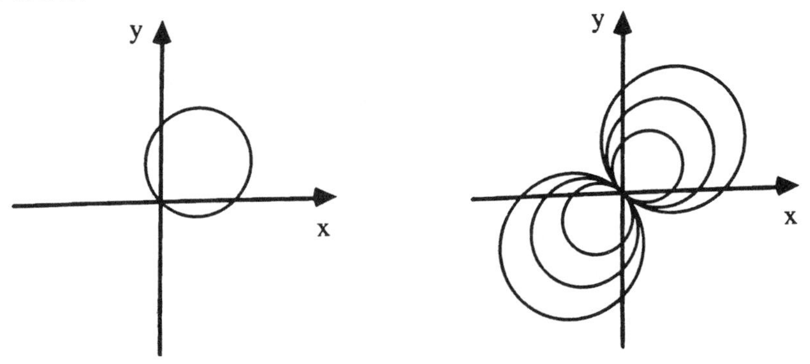

6. We want to show that $\boldsymbol{\sigma}'(t) = \mathbf{F}(\boldsymbol{\sigma}(t))$. Differentiate to get $\boldsymbol{\sigma}'(t) = (2e^{2t}, 1/t, -1/t^2)$. Calculate $\mathbf{F}(\boldsymbol{\sigma}(t)) = (2x, z, -z^2) = (2e^{2t}, 1/t, -(1/t)^2)$, which is indeed $\boldsymbol{\sigma}'(t)$.

3.4 Divergence and Curl of a Vector Field

GOALS

1. Given any vector field, be able to compute its divergence.

2. Given any vector field in $\mathbf{R}^3$, be able to compute its curl.

3. Be able to explain the physical significance of the divergence and the curl.

STUDY HINTS

1. *The operation* ∇. This operator, called "del", tells you to assemble the vector of partial derivatives: $(\partial/\partial x, \partial/\partial y, \partial/\partial z)$.

2. *Curl.* Note that the curl is a vector, not a scalar. Know the two notations: $\nabla \times \mathbf{F} = \text{curl}\,\mathbf{F}$. You should know that the curl is associated with rotations and that the term irrotational means that $\text{curl}\,\mathbf{F} = \mathbf{0}$.

3. *Valid space for curl.* Note that curl is a property of $\mathbf{R}^3$. One can not take the curl of vectors in dimensions higher than three. The two-dimensional vector $\mathbf{i} + \mathbf{j}$ is taken to mean $\mathbf{i} + \mathbf{j} + 0\mathbf{k}$ if a curl is desired.

4. *Divergence.* Note that the divergence is a scalar, not a vector. Know the two notations: $\nabla \cdot \mathbf{F} = \text{div}\,\mathbf{F}$. You should know that the divergence is a rate of expansion or compression. The term incompressible means that $\text{div}\,\mathbf{F} = 0$.

5. *Laplacian.* $\nabla^2 f$ means $\nabla \cdot (\nabla f)$, which is a scalar.

6. *Theorems.* The facts that $\nabla \times (\nabla f) = \mathbf{0}$ and $\nabla \cdot (\nabla \times \mathbf{F}) = 0$ are useful; it is nice to commit these to memory. If you need to know one of these facts and you forget them, you can always do the computation.

SOLUTIONS TO SELECTED EXERCISES

1. (b) The curl is $\nabla \times \mathbf{F} =$

$$\begin{vmatrix} \mathbf{i} & \mathbf{j} & \mathbf{k} \\ \partial/\partial x & \partial/\partial y & \partial/\partial z \\ yz & xz & xy \end{vmatrix} = \mathbf{i}\left(\frac{\partial}{\partial y}(xy) - \frac{\partial}{\partial y}(xz)\right) - \mathbf{j}\left(\frac{\partial}{\partial x}(xy) - \frac{\partial}{\partial z}(yz)\right) +$$

$$\mathbf{k}\left(\frac{\partial}{\partial x}(xz) - \frac{\partial}{\partial y}(yz)\right)$$

$$= (x - x)\mathbf{i} - (y - y)\mathbf{j} + (z - z)\mathbf{k} = \mathbf{0}\ .$$

2. (b) The divergence is $\nabla \cdot \mathbf{F} = \frac{\partial}{\partial x}(yz) + \frac{\partial}{\partial y}(xz) + \frac{\partial}{\partial z}(xy) = 0$.

3. (b) First, we compute that $\nabla f = (y + z)\mathbf{i} + (x + z)\mathbf{j} + (y + x)\mathbf{k}$. Then

$$\nabla \times \nabla f = \begin{vmatrix} \mathbf{i} & \mathbf{j} & \mathbf{k} \\ \partial/\partial x & \partial/\partial y & \partial/\partial z \\ y+z & x+z & y+x \end{vmatrix}$$

$$= \left[\frac{\partial}{\partial y}(y + x) - \frac{\partial}{\partial z}(x + z)\right]\mathbf{i} - \left[\frac{\partial}{\partial x}(y + x) - \frac{\partial}{\partial z}(y + z)\right]\mathbf{j} + \left[\frac{\partial}{\partial x}(x + z) - \frac{\partial}{\partial y}(y + z)\right]\mathbf{k}$$
$$= (1 - 1)\mathbf{i} - (1 - 1)\mathbf{j} + (1 - 1)\mathbf{k} = \mathbf{0}.$$

6. (a) $$\text{curl } \mathbf{F} = \nabla \times \mathbf{F} = \begin{vmatrix} \mathbf{i} & \mathbf{j} & \mathbf{k} \\ \partial/\partial x & \partial/\partial y & \partial/\partial z \\ 3x^2y & x^3 + y^3 & 0 \end{vmatrix} = \mathbf{k}\left(\frac{\partial}{\partial x}(x^3 + y^3) - \frac{\partial}{\partial y}(3x^2y)\right)$$
$$= \mathbf{k}\,(3x^2 - 3x^2) = \mathbf{0}.$$

(b) Note that if $\mathbf{F} = \nabla f$, then $3x^2y = (\partial/\partial x)\, f(x, y)$ and $(x^3 + y^3) = (\partial/\partial y)\, f(x, y)$. Integrate each equation:

$$f(x, y) = \int 3x^2y \; dx = x^3y + g(y),$$

where g is a function of y only, and

$$f(x, y) = \int (x^3 + y^3)\, dy = x^3y + \frac{y^4}{4} + h(x),$$

where h is a function of x only. In both cases, $f(x, y)$ must be the same, so compare both sides. If we let $g(y) = y^4/4$ and $h(x)$ be an arbitrary constant, then $f(x, y) = x^3y + y^4/4 + C$ satisfies $\nabla f = \mathbf{F}$.

8. (c) Let $\mathbf{F}(x, y) = e^x\cos y\, \mathbf{i} - e^x\sin y\, \mathbf{j}$, where the $\mathbf{i}$ component is the real part of $e^{x - iy}$, and the $\mathbf{j}$ component is the imaginary part. We calculate div $\mathbf{F}$ = $(\partial/\partial x)(e^x \cos y) + (\partial/\partial y)(-e^x \sin y) = e^x \cos y - e^x \cos y = 0$ and

$$\text{curl } \mathbf{F} = \begin{vmatrix} \mathbf{i} & \mathbf{j} & \mathbf{k} \\ \partial/\partial x & \partial/\partial y & \partial/\partial z \\ e^x\cos y & -e^x\sin y & 0 \end{vmatrix} = \mathbf{k}\left(\frac{\partial}{\partial x}(-e^x\sin y) - \frac{\partial}{\partial y}(e^x\cos y)\right)$$
$$= \mathbf{k}\,(-e^x\sin y + e^x\sin y) = \mathbf{0}.$$

So $\mathbf{F}$ is both incompressible and irrotational.

11. By the product rule,

$$\left.\frac{d}{dt}(\mathbf{v}\cdot\mathbf{w})\right|_{t=0} = \frac{d\mathbf{v}}{dt}\cdot\mathbf{w} + \mathbf{v}\cdot\frac{d\mathbf{w}}{dt}$$

$$= [\mathbf{D_x F}(0)\,\mathbf{v}]\cdot\mathbf{w} + \mathbf{v}\cdot[\mathbf{D_x F}(0)\,\mathbf{w}]$$

Denote matrix $\mathbf{D_x F}(0)$ by $\mathbf{A}$. We shall first show that $\mathbf{A}^T\mathbf{v}\cdot\mathbf{w} = \mathbf{v}\cdot\mathbf{Aw}$. Recall (from linear algebra) that $\mathbf{v}\cdot\mathbf{w} = \mathbf{w}^T\mathbf{v}$ for two vectors $\mathbf{v}$ and $\mathbf{w}$, and that $(\mathbf{AB})^T = \mathbf{B}^T\mathbf{A}^T$. Thus,

$$\mathbf{v}\cdot\mathbf{Aw} = (\mathbf{Aw})^T\,\mathbf{v} = \mathbf{w}^T(\mathbf{A}^T\mathbf{v}) = \mathbf{A}^T\mathbf{v}\cdot\mathbf{w}.$$

From this and the distributivity property, we conclude that

$$\left.\frac{d}{dt}(\mathbf{v}\cdot\mathbf{w})\right|_{t=0} = [(\mathbf{D_x F}(0) + (\mathbf{D_x F}(0))^T)\,\mathbf{v}]\cdot\mathbf{w}.$$

3.5 Vector Differential Calculus

GOALS

1. Be able to manipulate expressions involving the cross product, the dot product and the del operator.

STUDY HINTS

1. *Review.* This section gives you the opportunity to review how to compute a gradient, a curl, a divergence, and a Laplacian.

2. *Formulas in* $\mathbf{R}^3$. Note that the cross product or curl occurs in the formulas of table 3.1; therefore, it is assumed that the formulas are used in $\mathbf{R}^3$ and not in higher dimensions.

3. *Importance of table 3.1.* These formulas are useful for developing the theory of vectors; however, you should not memorize the table. You can refer back to the table as needed. It will become obvious which formulas are most important as you refer back to them frequently.

4. *Exercise 8.* These formulas are referred to quite frequently in the examples. You should do this exercise even if it is not assigned.

5. *Cylindrical and spherical coordinates.* Many problems in physics and engineering involve objects of cylindrical and spherical shape. The corresponding coordinate systems are useful for dealing with them. The formulas for div, grad, and curl in these coordinates are normally not memorized, but be prepared to refer to them depending on the tastes of your instructor.

SOLUTIONS TO SELECTED EXERCISES

3. (9) We want to prove that $\operatorname{div}(\mathbf{F} \times \mathbf{G}) = \mathbf{G} \cdot \operatorname{curl} \mathbf{F} - \mathbf{F} \cdot \operatorname{curl} \mathbf{G}$. Let A_x denote $\partial A/\partial x$. Let

$$\mathbf{F} = (F_1, F_2, F_3) \text{ and } \mathbf{G} = (G_1, G_2, G_3).$$

Then
$$\begin{aligned}
\operatorname{div}(\mathbf{F} \times \mathbf{G}) &= \operatorname{div}[(F_2G_3 - F_3G_2)\mathbf{i} + (G_1F_3 - F_1G_3)\mathbf{j} + (F_1G_2 - F_2G_1)\mathbf{k}] \\
&= (F_2G_3 - F_3G_2)_x + (G_1F_3 - F_1G_3)_y + (F_1G_2 - F_2G_1)_z \\
&= (F_2)_xG_3 + F_2(G_3)_x - (F_3)_xG_2 - F_3(G_2)_x + \\
&\quad (G_1)_yF_3 + G_1(F_3)_y - (F_1)_yG_3 - F_1(G_3)_y + \\
&\quad (F_1)_zG_2 + F_1(G_2)_z - (F_2)_zG_1 - F_2(G_1)_z \\
&= \{G_1[(F_3)_y - (F_2)_z] - G_2[(F_3)_x - (F_1)_z] + \\
&\quad G_3[(F_2)_x - (F_1)_y]\} - \{F_1[(G_3)_y - (G_2)_z] - \\
&\quad F_2[(G_3)_x - (G_1)_z] + F_3[(G_2)_x - (G_1)_y]\} \\
&= \mathbf{G} \cdot \operatorname{curl} \mathbf{F} - \mathbf{F} \cdot \operatorname{curl} \mathbf{G}.
\end{aligned}$$

5. (c) By substituting directly,

$$\begin{aligned}
(\mathbf{F} \cdot \nabla)\mathbf{G} &= [2xz^2(\partial/\partial x) + (\partial/\partial y) + y^3zx(\partial/\partial z)](x^2\mathbf{i} + y^2\mathbf{j} + z^2\mathbf{k}) \\
&= 2xz^2(\partial/\partial x)(x^2\mathbf{i} + y^2\mathbf{j} + z^2\mathbf{k}) + (\partial/\partial y)(x^2\mathbf{i} + y^2\mathbf{j} + z^2\mathbf{k}) + \\
&\quad y^3zx(\partial/\partial z)(x^2\mathbf{i} + y^2\mathbf{j} + z^2\mathbf{k}) \\
&= 4xz^2\mathbf{i} + 2y\mathbf{j} + 2y^3zx\mathbf{k} = 4x^2z^2\mathbf{i} + 2y\mathbf{j} + 2y^3z^2x\mathbf{k}.
\end{aligned}$$

8. (a) We have $\nabla(1/r) = \nabla(1/\sqrt{x^2 + y^2 + z^2})$. Begin by finding $(\partial/\partial x)(1/r)$ or

$$\frac{\partial}{\partial x}\left(\frac{1}{\sqrt{x^2 + y^2 + z^2}}\right) = \frac{-2x}{2(x^2 + y^2 + z^2)^{3/2}} = \frac{-x}{(x^2 + y^2 + z^2)^{3/2}} = \frac{-x}{r^3}.$$

By symmetry, $(\partial/\partial y)(1/r) = -y/r^3$ and $(\partial/\partial z)(1/r) = -z/r^3$. Therefore, $\nabla(1/r) = -(x\mathbf{i} + y\mathbf{j} + z\mathbf{k})/r^3 = -\mathbf{r}/r^3$. In general,

$$\nabla(r^n) = \nabla((x^2 + y^2 + z^2)^{n/2})$$

and

$$(\partial/\partial x)((x^2 + y^2 + z^2)^{n/2}) = (n/2)((x^2 + y^2 + z^2)^{n/2 - 1})2x = nxr^{n-2},$$

so by symmetry,

$$\nabla(r^n) = nr^{n-2}(x\mathbf{i} + y\mathbf{j} + z\mathbf{k}) = nr^{n-2}\mathbf{r}.$$

Finally,

$$\nabla(\log r) = \nabla\left(\log\left(\sqrt{x^2 + y^2 + z^2}\right)\right),$$

and

$$\frac{\partial}{\partial x}\left(\log\left(\sqrt{x^2 + y^2 + z^2}\right)\right) = \frac{1}{\sqrt{x^2 + y^2 + z^2}} \cdot \frac{\partial}{\partial x}\left(\sqrt{x^2 + y^2 + z^2}\right) = \frac{x}{r^2},$$

so by symmetry, $\nabla(\log r) = (x\mathbf{i} + y\mathbf{j} + z\mathbf{k})/r^2 = \mathbf{r}/r^2$.

(b) Using the results of part (a), $\nabla^2(1/r) = \nabla \cdot \nabla(1/r) = \nabla \cdot (-\mathbf{r}/r^3) =$

$$-\left[\frac{\partial}{\partial x}\left(\frac{x}{\left(\sqrt{x^2+y^2+z^2}\right)^3}\right)+\frac{\partial}{\partial y}\left(\frac{y}{\left(\sqrt{x^2+y^2+z^2}\right)^3}\right)+\frac{\partial}{\partial z}\left(\frac{z}{\left(\sqrt{x^2+y^2+z^2}\right)^3}\right)\right].$$

The first partial derivative is

$$\frac{\partial}{\partial x}\left(\frac{x}{(x^2+y^2+z^2)^{3/2}}\right)=\frac{(x^2+y^2+z^2)^{3/2}-(3/2)\sqrt{x^2+y^2+z^2}\cdot 2x^2}{(x^2+y^2+z^2)^3}$$

$$=\frac{x^2+y^2+z^2-3x^2}{(x^2+y^2+z^2)^{5/2}}.$$

By symmetry,

$$\frac{\partial}{\partial y}\left(\frac{y}{(x^2+y^2+z^2)^{3/2}}\right)=\frac{x^2+y^2+z^2-3y^2}{(x^2+y^2+z^2)^{5/2}} \quad \text{and}$$

$$\frac{\partial}{\partial z}\left(\frac{z}{(x^2+y^2+z^2)^{3/2}}\right)=\frac{x^2+y^2+z^2-3z^2}{(x^2+y^2+z^2)^{5/2}}.$$

Then

$$\nabla\cdot\left(\frac{-\mathbf{r}}{r^3}\right)=\frac{3(x^2+y^2+z^2)-3(x^2+y^2+z^2)}{(x^2+y^2+z^2)^{5/2}}=0.$$

Similarly, part (a) tells us that in the general case,

$$\nabla^2(r^n)=\nabla\cdot\nabla(r^n)=\nabla\cdot(nr^{n-2}\mathbf{r}).$$

We compute

$$\frac{\partial}{\partial x}((x^2+y^2+z^2)^{n/2-1}\cdot x)$$

$$=n\left[(x^2+y^2+z^2)^{n/2-1}+2x^2\left(\frac{n}{2}-1\right)(x^2+y^2+z^2)^{n/2-2}\right].$$

Again, by symmetry,

$$\nabla\cdot(nr^{n-2}\mathbf{r})=n[3(x^2+y^2+z^2)^{n/2-1}+$$
$$(x^2+y^2+z^2)(n-2)(x^2+y^2+z^2)^{n/2-2}]$$
$$=nr^{n-2}(3+n-2)=n(n+1)r^{n-2}.$$

(c) The identity $\nabla\cdot(\mathbf{r}/r^3)=0$ follows immediately from part (b).

For the general case, use part (b) again to compute $\nabla\cdot(r^k\mathbf{r})$. Note that we only need to divide the general result by n and change $n-2$ to k (Why?). Then

$\nabla\cdot(r^k\mathbf{r})=(k+3)r^k=(n+3)r^n$, so changing the name of k to n gives the result.

(d) By a direct computation, we get

$$\nabla\times\mathbf{r}=\begin{vmatrix}\mathbf{i} & \mathbf{j} & \mathbf{k}\\ \partial/\partial x & \partial/\partial y & \partial/\partial z\\ x & y & z\end{vmatrix}=\mathbf{0}.$$

By formula 11 of table 3.1, the calculation just completed, and the general case in part (a),

$$\nabla \times (r^n\mathbf{r}) = r^n(\nabla \times \mathbf{r}) + \nabla(r^n) \times \mathbf{r}$$
$$= \mathbf{0} + nr^{n-2}(\mathbf{r} \times \mathbf{r}) = \mathbf{0} \ .$$

11. Let $x = r\cos\theta$ and $y = r\sin\theta$, so $\partial x/\partial r = \cos\theta$, $\partial y/\partial r = \sin\theta$, $\partial x/\partial\theta = -r\sin\theta$, and $\partial y/\partial\theta = r\cos\theta$. By the chain rule:

$$\frac{\partial u}{\partial r} = \frac{\partial u}{\partial x}\bullet\frac{\partial x}{\partial r} + \frac{\partial u}{\partial y}\bullet\frac{\partial y}{\partial r} = \cos\theta\left(\frac{\partial u}{\partial x}\right) + \sin\theta\left(\frac{\partial u}{\partial y}\right) \quad \text{and}$$
$$\frac{\partial u}{\partial \theta} = \frac{\partial u}{\partial x}\bullet\frac{\partial x}{\partial \theta} + \frac{\partial u}{\partial y}\bullet\frac{\partial y}{\partial \theta} = -r\sin\theta\left(\frac{\partial u}{\partial x}\right) + r\cos\theta\left(\frac{\partial u}{\partial y}\right).$$

Now solve for $\partial u/\partial y$: Multiply the expression for $\partial u/\partial r$ by $r\sin\theta$ and multiply the expression for $\partial u/\partial\theta$ by $\cos\theta$. Add them together:

$$r\sin\theta\left(\frac{\partial u}{\partial r}\right) + \cos\theta\left(\frac{\partial u}{\partial \theta}\right) = r\left(\frac{\partial u}{\partial y}\right) \text{ implies } \frac{\partial u}{\partial y} = \sin\theta\left(\frac{\partial u}{\partial r}\right) + \frac{1}{r}\cos\theta\left(\frac{\partial u}{\partial \theta}\right).$$

Similarly, we solve for $\partial u/\partial x$. Multiply the expression for $\partial u/\partial r$ by $r\cos\theta$ and multiply the expression for $\partial u/\partial\theta$ by $-\sin\theta$. Add them together:

$$r\cos\theta\left(\frac{\partial u}{\partial r}\right) - \sin\theta\left(\frac{\partial u}{\partial \theta}\right) = r\left(\frac{\partial u}{\partial x}\right) \text{ implies } \frac{\partial u}{\partial x} = \cos\theta\left(\frac{\partial u}{\partial r}\right) - \frac{1}{r}\sin\theta\left(\frac{\partial u}{\partial \theta}\right).$$

Use the chain rule again to find the second partial:

$$\frac{\partial^2 u}{\partial x^2} = \frac{\partial}{\partial x}\left(\frac{\partial u}{\partial x}\right) = \frac{\partial}{\partial r}\left(\frac{\partial u}{\partial x}\right)\bullet\frac{\partial r}{\partial x} + \frac{\partial}{\partial \theta}\left(\frac{\partial u}{\partial x}\right)\bullet\frac{\partial \theta}{\partial x}$$
$$= \left(\cos\theta\frac{\partial^2 u}{\partial r^2} + \frac{1}{r^2}\sin\theta\frac{\partial u}{\partial \theta} - \frac{1}{r}\sin\theta\frac{\partial^2 u}{\partial\theta\,\partial r}\right)\bullet\left(\frac{1}{\cos\theta}\right) +$$
$$\left(-\sin\theta\frac{\partial u}{\partial r} + \cos\theta\frac{\partial^2 u}{\partial\theta\,\partial r} - \frac{1}{r}\cos\theta\frac{\partial u}{\partial \theta} - \frac{1}{r}\sin\theta\frac{\partial^2 u}{\partial \theta^2}\right)\bullet\left(\frac{-1}{r\sin\theta}\right)$$
$$= \frac{\partial^2 u}{\partial r^2} + \frac{1}{r^2}\frac{\partial u}{\partial \theta}\frac{\sin\theta}{\cos\theta} - \frac{1}{r}\frac{\partial^2 u}{\partial\theta\,\partial r}\frac{\sin\theta}{\cos\theta} +$$
$$\frac{1}{r}\frac{\partial u}{\partial r} - \frac{1}{r}\frac{\partial^2 u}{\partial\theta\,\partial r}\frac{\cos\theta}{\sin\theta} + \frac{1}{r^2}\frac{\cos\theta}{\sin\theta}\frac{\partial u}{\partial \theta} + \frac{1}{r^2}\frac{\partial^2 u}{\partial \theta^2}.$$

Similarly,

$$\frac{\partial^2 u}{\partial y^2} = \frac{\partial}{\partial y}\left(\frac{\partial u}{\partial y}\right) = \frac{\partial}{\partial r}\left(\frac{\partial u}{\partial y}\right)\bullet\frac{\partial r}{\partial y} + \frac{\partial}{\partial \theta}\left(\frac{\partial u}{\partial y}\right)\bullet\frac{\partial \theta}{\partial y} =$$
$$= \left(\frac{1}{\sin\theta}\right)\bullet\left(\sin\theta\frac{\partial^2 u}{\partial r^2} - \frac{1}{r^2}\cos\theta\frac{\partial u}{\partial \theta} + \frac{1}{r}\cos\theta\frac{\partial^2 u}{\partial r\,\partial\theta}\right) +$$
$$\left(\frac{1}{r\cos\theta}\right)\bullet\left(\cos\theta\frac{\partial u}{\partial r} + \sin\theta\frac{\partial^2 u}{\partial\theta\,\partial r} - \frac{1}{r}\cos\theta\frac{\partial u}{\partial \theta} - \frac{1}{r}\sin\theta\frac{\partial u}{\partial \theta}\right)$$

$$= \frac{\partial^2 u}{\partial r^2} - \frac{1}{r^2}\frac{\cos\theta}{\sin\theta}\frac{\partial u}{\partial\theta} + \frac{1}{r}\frac{\partial^2 u}{\partial r\,\partial\theta}\frac{\cos\theta}{\sin\theta} +$$

$$\frac{1}{r}\frac{\partial u}{\partial r} + \frac{1}{r}\frac{\sin\theta}{\cos\theta}\frac{\partial^2 u}{\partial r\,\partial\theta} + \frac{1}{r^2}\frac{\partial^2 u}{\partial\theta^2} - \frac{1}{r^2}\frac{\sin\theta}{\cos\theta}\frac{\partial u}{\partial\theta} .$$

Add them together and cancel the terms. We get

$$\frac{\partial^2 u}{\partial x^2} + \frac{\partial^2 u}{\partial y^2} = 2\left(\frac{\partial^2 u}{\partial r^2} + \frac{1}{r}\frac{\partial u}{\partial r} + \frac{1}{r^2}\frac{\partial^2 u}{\partial\theta^2}\right).$$

But since $\nabla^2 u = 0$ is given, this reduces to

$$\frac{\partial^2 u}{\partial r^2} + \frac{1}{r}\frac{\partial u}{\partial r} + \frac{1}{r^2}\frac{\partial^2 u}{\partial\theta^2} = 0 .$$

3.R Review Exercises for Chapter 3

SOLUTIONS TO SELECTED EXERCISES

1. (b) The divergence is $\nabla \cdot \mathbf{F} = \partial y/\partial x + \partial z/\partial y + \partial x/\partial z = 0$.

2. (b) The curl is $\nabla \times \mathbf{F} = \begin{vmatrix} \mathbf{i} & \mathbf{j} & \mathbf{k} \\ \partial/\partial x & \partial/\partial y & \partial/\partial z \\ y & z & x \end{vmatrix}$

$$= \left(\frac{\partial x}{\partial y} - \frac{\partial z}{\partial z}\right)\mathbf{i} + \left(\frac{\partial y}{\partial z} - \frac{\partial x}{\partial x}\right)\mathbf{j} + \left(\frac{\partial z}{\partial x} - \frac{\partial y}{\partial y}\right)\mathbf{k}$$

$$= -\mathbf{i} - \mathbf{j} - \mathbf{k} .$$

4. (b) We compute $\operatorname{div}\mathbf{F} = \nabla \cdot \mathbf{F} = \frac{\partial}{\partial x}(x^2) + \frac{\partial}{\partial y}(y^2) + \frac{\partial}{\partial z}(x^2) = 2x + 2y + 2z$.

Also,

$$\operatorname{curl}\mathbf{F} = \nabla \times \mathbf{F} = \begin{vmatrix} \mathbf{i} & \mathbf{j} & \mathbf{k} \\ \partial/\partial x & \partial/\partial y & \partial/\partial z \\ x^2 & y^2 & z^2 \end{vmatrix}$$

$$= \left[\frac{\partial}{\partial y}(z^2) - \frac{\partial}{\partial z}(y^2)\right]\mathbf{i} - \left[\frac{\partial}{\partial x}(z^2) - \frac{\partial}{\partial z}(x^2)\right]\mathbf{j} + \left[\frac{\partial}{\partial x}(y^2) - \frac{\partial}{\partial y}(x^2)\right]\mathbf{k} = \mathbf{0} .$$

5. (b) The velocity vector is $\mathbf{v}(t) = \boldsymbol{\sigma}'(t) = 2t\mathbf{i} + (-2t\sin(t^2))\mathbf{j} + 4t^3\mathbf{k}$. The acceleration vector is $\mathbf{a}(t) = \boldsymbol{\sigma}''(t) = 2\mathbf{i} + [-2\sin(t^2) - 4t^2\cos(t^2)]\mathbf{j} + 12t^2\mathbf{k}$. When $t = \sqrt{\pi}$, we have $\mathbf{v}(\sqrt{\pi}) = 2\sqrt{\pi}\,\mathbf{i} + 4\pi\sqrt{\pi}\,\mathbf{k}$ and $\mathbf{a}(\sqrt{\pi}) = 2\mathbf{i} - 4\pi\mathbf{j} + 12\pi\mathbf{k}$. The

speed is the length of the velocity vector. When $t = \sqrt{\pi}$, the speed is $[4\pi + 16\pi^3]^{1/2} = 2\sqrt{\pi + 4\pi^3}$. The equation of the tangent line is given by $\mathbf{l}(t) = \boldsymbol{\sigma}(\sqrt{\pi}) + t\,\mathbf{v}(\sqrt{\pi}) = (\pi - 1, -1, \pi^2) + t(2\sqrt{\pi}, 0, 4\pi\sqrt{\pi})$.

9. We use the equation $\mathbf{F} = m\mathbf{a} = m\boldsymbol{\sigma}''$. Here, $\boldsymbol{\sigma}'' = (2, -\sin t, -\cos t)$. At $t = 0$, $\mathbf{a}(0) = (2, 0, -1)$, so $\mathbf{F}(0) = m(2, 0, -1)$.

11. When $t = \pi$, the particle is at $(\pi, \pi^2, -\pi)$. At this point, the velocity vector is $\mathbf{v}(t) = \mathbf{c}'(t) = (1, 2t, \cos t - t \sin t)$, which is $(1, 2\pi, -1)$ at $t = \pi$. The equation of the tangent line is $\mathbf{l}(\lambda) = (\pi, \pi^2, -\pi) + \lambda(1, 2\pi, -1)$. We use $\lambda = \pi$ to find the particle's location because $\Delta t = 2\pi - \pi = \pi$. At $t = 2\pi$, the particle is at $(2\pi, 3\pi^2, -2\pi)$.

15. We want to show that $\boldsymbol{\sigma}'(t) = \mathbf{F}(\boldsymbol{\sigma}(t))$. Here, $x = 1/(1-t)$, $y = 0$, and $z = e^t/(1-t)$. We get $dx/dt = 1/(1-t)^2$, $dy/dt = 0$, and $dz/dt = (2e^t - te^t)/(1-t)^2$. Thus, $dx/dt = x^2$, $dy/dt = 0$, and $dz/dt = z(1 + x) = [e^t/(1-t)][1 + 1/(1-t)] = (2e^t - te^t)/(1-t)^2$.

16. From $x^2 = y^3 = z^5$, we get $y = x^{2/3}$ and $z = x^{2/5}$. Let $x = t$, so the path of the curve is $\boldsymbol{\sigma}(t) = (t, t^{2/3}, t^{2/5})$. Here, we have $1 \le t \le 4$ and $\boldsymbol{\sigma}'(t) = (1, (2/3)t^{-1/3}, (2/5)t^{-3/5})$ The arc length is

$$\int_1^4 \|\boldsymbol{\sigma}'(t)\|\,dt = \int_1^4 \sqrt{1 + \frac{4}{9}t^{-2/3} + \frac{4}{25}t^{-6/5}}\,dt\,.$$

CHAPTER 4
HIGHER-ORDER DERIVATIVES; MAXIMA AND MINIMA

4.1 Taylor's Theorem

GOALS

1. Be able to write down the first few terms of Taylor's formula for a given function.

STUDY HINTS

1. *Notation.* The summation symbol $\sum_{i,j=1}^{3}$ means to sum all possible combinations of (i, j) with i and j ranging from 1 to 3 , i.e., (i, j) = (1, 1), (1, 2), (1, 3), (2, 1), (2, 2), (2, 3), (3, 1), (3, 2), and (3, 3) . In general, if m indices are summed from 1 to n , there will be n^m terms; in our case there are $3^2 = 9$ terms.

2. *Review.* Before continuing, you may wish to review Taylor's formula from your one-variable calculus text. Recall that the Taylor series can be used to approximate the values of functions.

3. *Taylor's formula.* You should know the pattern for the general formula. As a reminder, Taylor's formula is

$$f(\mathbf{x}_0 + \mathbf{h}) = f(\mathbf{x}_0) + \sum_{i=1}^{n} h_i \frac{\partial f}{\partial x_i}(\mathbf{x}_0) +$$

$$\frac{1}{2!} \sum_{i,\,j=1}^{n} h_i h_j \frac{\partial^2 f}{\partial x_i \partial x_j}(\mathbf{x}_0) + \frac{1}{3!} \sum_{i,j,k=1}^{n} h_i h_j h_k \frac{\partial^3 f}{\partial x_i \partial x_j \partial x_k}(\mathbf{x}_0) + \dots .$$

The second term, which involves the second partial derivatives, will become important in the coming sections. The term involving the third partials sums up 3^n terms, so it may be unreasonable to ask you to compute all of these terms unless $n = 2$.

4. *Computing Taylor's formula.* Remember that you will need to compute *all* of the partial derivatives of the same order. For example, when computing second partials, one must compute $\frac{\partial^2 f}{\partial x_1^2}, \frac{\partial^2 f}{\partial x_2^2}, \dots, \frac{\partial^2 f}{\partial x_n^2}$ as well as *all* of the mixed partials $\frac{\partial^2 f}{\partial x_1 \partial x_2}$, $\frac{\partial^2 f}{\partial x_2 \partial x_1}$, $\frac{\partial^2 f}{\partial x_1 \partial x_3}$, etc. Note that we do not need to compute $\frac{\partial^2 f}{\partial x_i \partial x_j}$, $i \neq j$ twice

because mixed partials are equal.

5. *Taylor's formula remainder.* Recall that in one-variable calculus, the remainder is determined at some point between x_0 and $x_0 + h$. Now, the remainder is determined at some point on the *line* between the vectors $\mathbf{x}_0$ and $\mathbf{x}_0 + \mathbf{h}$, where $\mathbf{h} = (h_1, h_2, \ldots, h_n)$.

SOLUTIONS TO SELECTED EXERCISES

1. Recall that Taylor's formula is a polynomial which approximates a function. If our function itself is a polynomial, then this function must be its own Taylor series as well. Hence, the second-order Taylor formula for f is $f(h_1, h_2) = (h_1 + h_2)^2 = h_1^2 + 2h_1h_2 + h_2^2$ at $x_0 = 0$, $y_0 = 0$.
   Alternatively, $\partial f/\partial x = \partial f/\partial y = 2(x + y)$ and $\partial^2f/\partial x^2 = \partial^2f/\partial x\partial y = \partial^2f/\partial y\partial x = \partial^2f/\partial y^2 = 2$. At $(0, 0)$, $f(x, y)$ and the first partials are all 0, and all of the second partials are 2. Thus, the Taylor approximation is $(1/2)(2)(h_1^2 + 2h_1h_2 + h_2^2) = (h_1 + h_2)^2$.

5. Here, $f(0, 0) = 1$, $f_x = y \cos xy - y \sin xy$, $f_y = x \cos xy - x \sin xy$, $f_{xx} = -y^2 \sin xy - y^2 \cos xy$, $f_{xy} = f_{yx} = \cos xy - xy \sin xy - \sin xy - xy \cos xy$, and $f_{yy} = -x^2 \sin xy - x^2 \cos xy$. At $(0, 0)$, we have $f_x = f_y = f_{xx} = f_{yy} = 0$ and $f_{xy} = 1$. Thus the Taylor series is
$$f(h_1, h_2) = 1 + 0h_1 + 0h_2 + (1/2)(0h_1^2 + 2h_1h_2 + 0h_2^2) = 1 + h_1h_2 + R_2(h, 0).$$

7. (b) For $x > 0$ and $x < 0$, $f(x) = \exp(-1/x)$ is infinitely differentiable. For $x = 0$, we must use the definition of the derivative:
$$f'(0) = \lim_{x\to 0^+} \frac{f(x) - f(0)}{x} = \lim_{x\to 0^+} \frac{\exp(-1/x)}{x}.$$
Let $u = 1/x$. We must show that $\lim_{u\to\infty} u \exp(-u) = 0$. By l'Hôpital's rule,
$$\lim_{u\to\infty} ue^{-u} = \lim_{u\to\infty}\left(\frac{u}{e^u}\right) = \lim_{u\to\infty}\left(\frac{1}{e^u}\right) = 0.$$
In general,
$$f^{(n)}(0) = \lim_{x\to 0^+} \frac{f^{(n-1)}(x) - f^{(n-1)}(0)}{x}.$$
$$f''(x) = \frac{1}{x^2}\exp\left(\frac{-1}{x}\right),\ f^{(3)}(x) = \left(\frac{-2}{x^3} + \frac{1}{x^4}\right)\exp\left(\frac{-1}{x}\right),$$
so if we can show that $\lim_{u\to\infty} u^ne^{-u} = 0$ for all n, then we can conclude that f is C^∞ at $x = 0$ as well. Again, use l'Hôpital's rule n times:
$$\lim_{u\to\infty} u^ne^{-u} = \lim_{u\to\infty} \frac{u^n}{e^u} = \ldots = \lim_{u\to\infty} \frac{n!}{e^u} = 0.$$
So f is C^∞, with all derivatives equalling 0 at $x = 0$. Now, $f(0 + h) = \exp(-1/h) > 0$, but

$$f(0) + f'(0)h + \dots + \frac{f^{(k)}(0)}{k!} h^k = 0 .$$

Hence f is not analytic .

4.2 Extrema of Real-Valued Functions

GOALS

1. Be able to find the critical points of a real-valued two-variable function.
2. Be able to use the Hessian to classify the critical points of a function.

STUDY HINTS

1. *Definitions.* (a) A local or relative extrema is a point $\mathbf{x}_0$ where $f(\mathbf{x}_0)$ is largest or smallest in a small *neighborhood* of $\mathbf{x}_0$.
   (b) An absolute extrema is a point $\mathbf{x}_0$ where $f(\mathbf{x}_0)$ is largest or smallest on the entire domain.
   (c) Critical points occur where *all* first partial derivatives are zero.
   (d) A saddle point is a critical point which is *not* a local extrema.

2. *Critical point - extremum relationship.* All extrema occur at critical points, but not all critical points are extrema. Critical points may also be saddle points.

3. *Real-valued functions.* Note that we are comparing function values for real-valued functions, *not* vector-valued functions.

4. *Finding extrema.* If $\partial f/\partial x = 0$ and $\partial f/\partial y = 0$ have more than one solution, each combination of (x, y) which satisfies these conditions must be considered. You must be complete in your analysis. See example 7.

5. *Hessian.* This is denoted by $Hf(\mathbf{x}_0)(\mathbf{h})$ and it is equal to the second term of Taylor's formula, which is $\frac{1}{2!} \sum_{i,j=1}^{n} h_i h_j \frac{\partial^2 f}{\partial x_i \partial x_j}(\mathbf{x}_0)$.

6. *Determining definiteness.* To determine definiteness in general, we need to know the determinants of the diagonal submatrices of the Hessian matrix. Starting from the upper left-hand corner, i.e., a_{11}, compute the determinants. If they are all positive, then the Hessian is positive definite. If $a_{11} < 0$ and the signs alternate, then the Hessian is negative definite. Note that this test includes theorem 5.

7. *Usefulness of Hessian.* At a critical point, the first partial derivatives are all zero, so Taylor's formula reduces to

   $f(\mathbf{x}_0 + \mathbf{h}) = f(\mathbf{x}_0)$ + Hessian + remainder,

   where the remainder is small compared to the Hessian. Thus, if $Hf(\mathbf{x}_0)$ is positive definite, then

   $f(\mathbf{x}_0 + \mathbf{h}) = f(\mathbf{x}_0)$ + Hessian + remainder $< f(\mathbf{x}_0)$,

   so $f(\mathbf{x}_0)$ is a relative minimum. Similarly, if $Hf(\mathbf{x}_0)$ is negative definite, then

   $f(\mathbf{x}_0) > f(\mathbf{x}_0)$ + Hessian ,

   and $f(\mathbf{x}_0)$ is a relative maximum.

8. *Classifying a critical point* $\mathbf{x}_0$. (a) If $Hf(\mathbf{x}_0)$ is positive definite, then $\mathbf{x}_0$ is a local minimum.
   (b) If $Hf(\mathbf{x}_0)$ is negative definite, then $\mathbf{x}_0$ is a local maximum.
   (c) If $Hf(\mathbf{x}_0)$ does not satisfy (a) or (b) and not all of the submatrices are zero, then $\mathbf{x}_0$ is a saddle point.

9. *Minimizing distance.* Example 7 shows that distances may be minimized by analyzing d^2 . This is justified by the chain rule. By differentiating d^2 , the chain rule gives $2d(\partial d/\partial x)$ and $2d(\partial d/\partial y)$, so we again are solving $\partial d/\partial x = 0$ and $\partial d/\partial y = 0$. Since $d \geq 0$, the maximum (or minimum) of d^2 would be the maximum (or minimum) of d.

10. *Guaranteeing absolute extrema.* In $\mathbf{R}^n$, if a domain is closed and bounded, and f is continuous on the domain, then there exists an absolute minimum and an absolute maximum. All three conditions are necessary. Think about what happens if the domain is not closed, if the domain is unbounded, or if f is not continuous. Compare this to the extreme value theorem of one-variable calculus.

11. *Locating boundary extrema.* In this section, one parametrizes the given boundary of a region and then differentiates to determine the local minimum and maximum points. Another method is introduced in the next section.

SOLUTIONS TO SELECTED EXERCISES

3. We compute the partials: $\partial f/\partial x = 2x + 2y$ and $\partial f/\partial y = 2y + 2x$. These partials vanish at points such that $x = -y$, so these are the critical points. Since $f(x, y) = (x + y)^2 \geq 0$ for all (x, y) , the extrema must be minima.

5. By the chain rule, $\partial f/\partial x = 2x \exp(1 + x^2 - y^2)$ and $\partial f/\partial y = -2y \exp(1 + x^2 - y^2)$. Setting these partials equal to 0 , we find that (0, 0) is the only critical point. To classify this critical point, recall that the exponential function is monotonic (it keeps increasing or decreasing). Looking at the function along the x axis, set $y = 0$ to get $f(x, 0) = \exp(1 + x^2)$. Thus, $x = 0$ is obviously a minimum. If we look at the y axis, we set $x = 0$ to get $f(0, y) = \exp(1 - y^2)$. The reader should be convinced that $y = 0$ is a maximum. Therefore, we conclude that (0, 0) is a saddle point.

9. At (0, 0) , $\cos(x^2 + y^2) = 1$. The cosine function can not exceed 1 , so (0, 0) is a

maximum. At $(\sqrt{\pi/2}, \sqrt{\pi/2})$, $\cos(x^2 + y^2) = \cos(\pi) = -1$, so $(\sqrt{\pi/2}, \sqrt{\pi/2})$ is a minimum since the cosine function is never less than -1. Now, consider the critical point $(0, \sqrt{\pi})$. We calculate $\cos(x^2 + y^2) = -1$, so this critical point is also a minimum. Note: This exercise cannot be done by the discriminant test (theorem 5).

10. The partial derivatives are $\partial f/\partial x = \sin y$ and $\partial f/\partial y = 1 + x \cos y$. Setting these equal to 0, we see that the critical points are $(-1, 2n\pi)$ and $(1, (2n+1)\pi)$, where n is an integer. Since $\partial^2 f/\partial x^2 = 0$, we cannot use the second derivative test. Consider the case when $x = 1$. Our function becomes $f(y) = y + \sin y$. The graph of $f(y)$ shows that when $y = (2n + 1)\pi$, there is an inflection point there. Therefore, we conclude that the point $(1, (2n+1)\pi)$ is a saddle point since, along the line $x = 1$, the critical point is neither a minimum nor a maximum. Similarly, if $x = -1$, then $f(y) = y - \sin y$ has inflection points at $y = 2n\pi$. Therefore, the critical points $(-1, 2n\pi)$ are also inflection points.

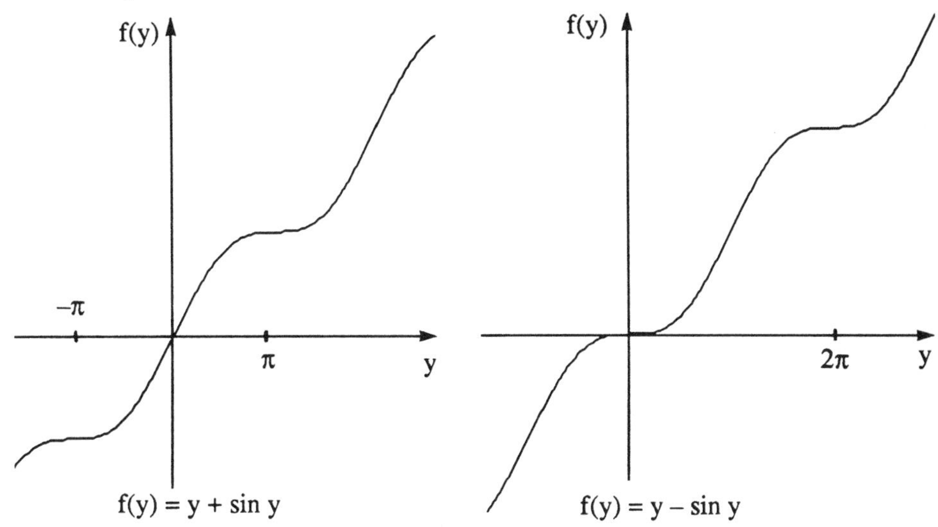

14. First, we compute $\partial f/\partial x = (y \cos xy)/(2 + \sin xy)$ and $\partial f/\partial y = (x \cos xy)/(2 + \sin xy)$. Since the denominator is between 1 and 3, we only need to solve $y \cos xy = 0$ and $x \cos xy = 0$. From $y \cos xy = 0$, we get $y = 0$ or $xy = (2n + 1)\pi/2$, where n is an integer. From $x \cos xy = 0$, we get the additional solution $x = 0$. To classify the extrema, we look at $f(x, y)$. Since $-1 \le \sin xy \le 1$, we have $\ln(1) \le f(x, y) \le \ln(3)$. So when $xy = -3\pi/2, \pi/2, \ldots, (4n + 1)\pi/2$, we have $f(x, y) = \ln(3)$ and these points are local maxima. Similarly, when $xy = -\pi/2, 3\pi/2, 7\pi/2, \ldots, (4n + 3)\pi/2$, we have $f(x, y) = \ln(1)$ and we have local minima. Now, look at $(x, y) = (0, 0)$. If $x = y$, then $\ln(2 + \sin x^2)$ is a minimum when $x = y = 0$. On the other hand, when $x = -y$, we have $\ln(2 + \sin(-x^2)) = \ln(2 - \sin x^2)$, which is a maximum at $x = y = 0$. Therefore, the point $(0, 0)$ is a saddle point (see figures on next page).

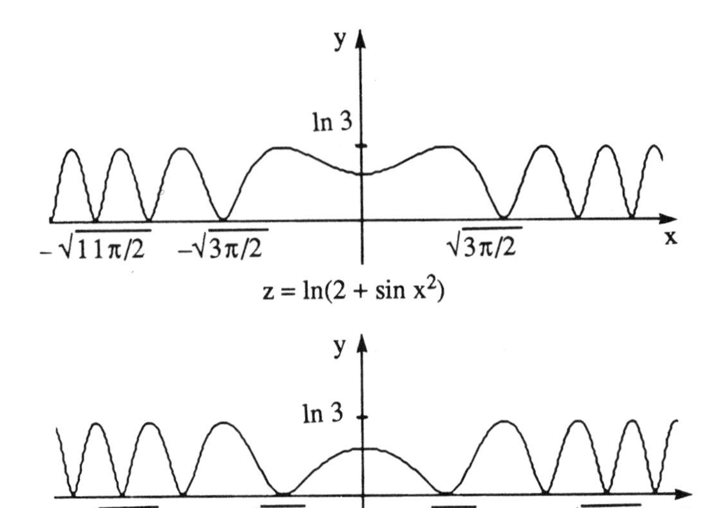

$z = \ln(2 + \sin x^2)$

$z = \ln(2 - \sin x^2)$

17. (a) We calculate the partials: $\partial f/\partial x = -6x(y - x^2) - 2x(y - 3x^2)$ and $\partial f/\partial y = (y - x^2) + (y - 3x^2)$. Setting $\partial f/\partial y$ equal to 0, we get $2y - 4x^2 = 0$, which implies $y = 2x^2$. Then substitution into $\partial f/\partial x$ gives $-6x(2x^2 - x^2) - 2x(2x^2 - 3x^2) = -6x^3 + 2x^3 = 0$, which implies that $x = 0$ and so $y = 0$, Therefore, $(0, 0)$ is a critical point.

    (b) Let $g(t) = (at, bt)$. Then $f(g(t)) = (bt - 3a^2t^2)(bt - a^2t^2) = b^2t^2 - 4a^2bt^3 + 3a^4t^4$. Differentiation gives $f'(g(t)) = 2tb^2 - 12a^2bt^2 + 12a^4t^3 = t(2b^2 - 12a^2bt + 12a^4t^2)$ which implies that $t = 0$ is one of the solutions. The second derivative is

    $f''(t) = 2b^2 - 24a^2bt + 36a^4t^2$ and $f''(0) = 2b^2 \geq 0$, independent of b (or a), except possibly at $b = 0$. For $b = 0$, $f(g(t)) = 3a^4t^4$, which has a minimum at $t = 0$. Thus $t = 0$ is a relative minimum of $f(t)$ along any straight line through the origin.

    (c) Look at the parabola $y = 2x^2$. Then $f(x, y) = (-x^2)(x^2) = -x^4 < 0$. So for this particular direction, $f(0, 0)$ is a maximum.

21. Given a volume V, suppose the dimensions are $x \times y \times z$, then $V = xyz$ and the surface area is $S = 2xy + 2yz + 2xz$. We want to minimize S. Solving for z, we get $z = V/xy$. Then $S = 2xy + 2V/x + 2V/y$. Taking partials, we get

    $$\partial S/\partial x = 2y - 2V(1/x^2) \qquad (1)$$
    $$\partial S/\partial y = 2x - 2V(1/y^2) \qquad (2)$$

    Setting (1) equal to 0, we get $y = V/x^2$. Substitute this into (2) and set it equal to 0: $2x - 2V/(V/x^2)^2 = 0$, which is equivalent to $x - x^4/V = 0$, or $x^3 = V$, or $x = V^{1/3}$. Then $y = V/x^2 = V/V^{2/3} = V^{1/3}$, and $z = V/xy = V/(V^{1/3}V^{1/3}) = V^{1/3}$. All three dimensions are equal, hence the box is a cube.

26. We compute the partial derivatives: $\partial f/\partial x = anx^{n-1}$ and $\partial f/\partial y = cny^{n-1}$. Setting these equal to 0, we see that the critical point is $(0, 0)$. The second partial derivatives are $\partial^2 f/\partial x^2 = n(n - 1)ax^{n-2}$, $\partial^2 f/\partial y^2 = n(n - 1)cy^{n-2}$, and $\partial^2 f/\partial x\partial y = 0$. By the second

derivative test, we conclude that the origin is a maximum if a and c are both negative and n is even. The origin is a minimum if both a and c are positive and n is even. For all other cases, we have a saddle point at the origin.

29. We want to find the extrema on the disk. First, check to see if an extrema occurs inside the disk. Take the partials of f and set them equal to 0:
$$\partial f/\partial x = 2x + y = 0$$
$$\partial f/\partial y = 2y + x = 0 .$$
So $(x, y) = (0, 0)$ is a relative extremum and $f(0,0) = 0$. On the boundary, we let $x = \cos\theta$, $y = \sin\theta$. Then $f(x, y) = f(\theta) = 1 + \sin\theta\cos\theta$. Differentiation gives $f'(\theta) = \cos^2\theta - \sin^2\theta$. Setting $f'(\theta) = 0$, we get $\cos^2\theta = \sin^2\theta$, which means that $\theta = \pi/4$, $3\pi/4$, $5\pi/4$, or $7\pi/4$ on the interval $0 \le \theta \le 2\pi$. For $\theta = \pi/4$, $f(x, y) = f(1/\sqrt{2}, 1/\sqrt{2}) = 1/2 + 1/2 + 1/2 = 3/2$; for $\theta = 3\pi/4$, $f(x, y) = f(-1/\sqrt{2}, 1/\sqrt{2}) = 1/2 - 1/2 + 1/2 = 1/2$; for $\theta = 5\pi/4$, $f(x,y) = f(-1/\sqrt{2}, -1/\sqrt{2}) = 1/2 + 1/2 + 1/2 = 3/2$; and for $\theta = 7\pi/4$, $f(x, y) = f(1/\sqrt{2}, 1/\sqrt{2}) = 1/2 - 1/2 + 1/2 = 1/2$. So the absolute maxima occurs at $(1/\sqrt{2}, 1/\sqrt{2})$ and $(-1/\sqrt{2}, -1/\sqrt{2})$, with the maximum value of 3/2, and the absolute minimum occurs at $(0, 0)$, with the minimum value of 0.

32. First, it is always wise to check for extrema inside the region. We set the partial derivatives equal to 0: $\partial f/\partial x = y = 0$ and $\partial f/\partial y = x = 0$. So $(0, 0)$ is a local extremum, and at $(0, 0)$, $f(x, y) = 0$. However, we have not checked the points on the boundary. For the line segment at $y = 1$, we have $-1 \le x \le 1$. On this line segment $f(x, y) = 1 \cdot x = x$, so the minimum for this line segment occurs at $(1, -1)$ and the maximum occurs at $(1, 1)$. Similarly, we can analyze the other three line segments. Therefore, at two of the corners, namely $(1, 1)$ and $(-1, -1)$, $f(x, y) = 1$ And at the other two corners, $f(x, y) = -1$. So $(0, 0)$ is not an absolute extremum; the maxima are at $(1, 1)$ and $(-1, -1)$ and the minima occur at $(1, -1)$ and $(-1, 1)$.

37. Suppose $\nabla^2 u = 0$ and u achieves its minimum on $D \backslash \partial D$. Let $u_n(x, y) = u(x, y) - (1/n)\, e^x$, then $\nabla^2 u_n = -(1/n)\, e^x < 0$. Thus, u_n is strickly superharmonic and by exercise 35, can have minimum only on ∂D, say, at $\mathbf{p}_n = (x_n, y_n)$. We have
$u_n(x_n, y_n) = u(x_n, y_n) - (1/n)\exp(x_n) \le u(x_n, y_n) - e^1/n$, since all $x_n \le 1$ on ∂D.
If (x_0, y_0) is a point in D then $u_n(x_0, y_0) > u_n(x_n, y_n)$, or
$$u(x_0, y_0) - e/n > u(x_n, y_n) - e/n .$$
Estimating the left-hand side upward (since $e/n > 0$), we have
$$u(x_0, y_0) > u(x_n, y_n) - e/n.$$
Since D is closed and bounded and $u(x, y)$ is continuous, there must be a point $\mathbf{q} = (x_\infty, y_\infty)$ on ∂D such that the above inequality holds in an arbitrarily small neighborhood of $\mathbf{q}$. Hence $u(x_0, y_0) \ge u(x_\infty, y_\infty)$ and u has a minimum on the boundary ∂D.

40. First, notice that $h = (y/2)\tan\theta$ and $d = (y/2)\sec\theta$. Thus, we want to maximize $A = xy + (1/2)yh = xy + (y^2/4)\tan\theta$. We are given $P = 2x + y + 2d = 2x + y + y\sec\theta$. Note that P is constant. Solving for x, we get $x = (1/2)(P - y - y\sec\theta)$ and our area function becomes $A(y, \theta) = (Py - y^2 - y^2\sec\theta)/2 + (y^2\tan\theta)/4$. Taking the partial derivatives, we get $\partial A/\partial y = P/2 - y - y\sec\theta + (y/2)\tan\theta$ and $\partial A/\partial\theta = (-y^2/2)\sec\theta\tan\theta + (y^2/4)\sec^2\theta = (y^2/4)\sec\theta(-2\tan\theta + \sec\theta)$. Setting $\partial A/\partial\theta = 0$ we get $y = 0$ or $2\tan\theta = \sec\theta$, i.e., $\sin\theta = 1/2$, i.e., $\theta = \pi/6$. The solution $y = 0$ is impossible for this geometric problem. When $\theta = \pi/6$, we have $\sec\theta = 2/\sqrt{3}$ and $\tan\theta = 1/\sqrt{3}$, so $\partial A/\partial y = P/2 - y - 2y/\sqrt{3} + y/2\sqrt{3}$. Setting $\partial A/\partial y = 0$, we get

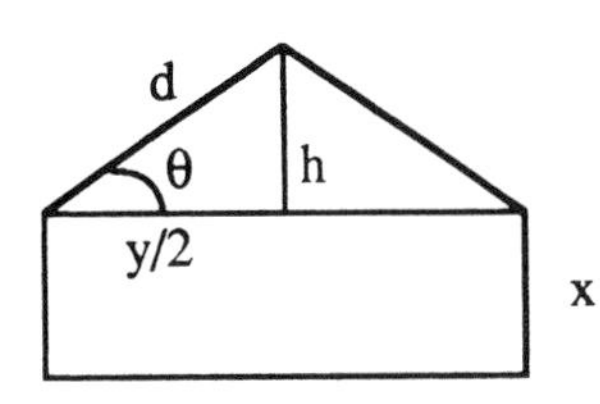

$$y = \frac{P}{2}\left(1 + \frac{2}{\sqrt{3}} - \frac{1}{2\sqrt{3}}\right)^{-1} = \frac{P\sqrt{3}}{2\sqrt{3} - 5}.$$

Therefore, the maximum area is

$$\frac{Py - y^2 - y^2\sec\theta}{2} + \frac{y^2\tan\theta}{4},$$

where

$$y = \frac{P\sqrt{3}}{2\sqrt{3} - 5} \quad \text{and} \quad \theta = \frac{\pi}{6}.$$

4.3 Constrained Extrema and Lagrange Multipliers

GOALS

1. Be able to find the extrema if one or more constraints are given.

2. Be able to analyze the critical points of a function with one constraint by using the bordered Hessian.

STUDY HINTS

1. *Notation.* The bar in $f \mid S$ means "restricted to". For example, if $g(x, y) = x + y$, then $g \mid (x = 2)$ means we want to consider the function $g = 2 + y$.

2. *Method for finding constrained extrema.* This is an alternative method which uses equations (3) rather than (2). First, rewrite the constraint so that the right-hand side is zero; for example, $x + y = 2$ becomes $x + y - 2 = 0$. Then consider the function

$h(\mathbf{x}, \lambda) = f(\mathbf{x}) + \lambda\cdot$(constraint). Solve $\partial h/\partial x_i = 0$ and $\partial h/\partial \lambda = 0$. In example 2, we analyze $h(\mathbf{x}, \lambda) = x^2 - y^2 + \lambda(x^2 + y^2 - 1)$ and in example 3, $h(\mathbf{x}, \lambda) = x + z + \lambda(x^2 + y^2 + z^2 - 1)$.

3. *Solving equations.* In general, we are not interested in the value of λ; only the values of the variables. Sometimes, solving λ in terms of the variables, thus eliminating λ, is the right thing to do.

4. *Cautions.* Remember, all of your equations must be solved simultaneously. Solving one equation alone does not aid in finding an extremum.

5. *Generalization.* If there is more than one constraint, then $\nabla f(\mathbf{x}_0) = \lambda_1 \nabla g_1(\mathbf{x}_0) + \lambda_2 \nabla g_2(\mathbf{x}_0) + \ldots + \lambda_n \nabla g_n(\mathbf{x}_0)$. The right-hand sides of equations (2) will have the form $\sum_{i=1}^{n} \lambda_i \frac{\partial g_i}{\partial x_j}$ in place of $\lambda \frac{\partial g}{\partial x_j}$ and there will be extra constraint equations. If you prefer to use equations (3), let $h(\mathbf{x}, \lambda) = f(\mathbf{x}) + \Sigma \lambda_i g_i(\mathbf{x})$ and solve $\partial h/\partial x_j = 0$ and $\partial h/\partial \lambda_i = 0$ simultaneously.

6. *Bordered Hessian.* This is the Hessian with a "border", which are the additional top row and the additional left column. The entries from left to right or top to bottom are $0, -\partial g/\partial x_1, -\partial g/\partial x_2, \ldots, -\partial g/\partial x_n$ where g is the constraint. Note that all entries except for the "border" are second partial derivatives.

7. *Classifying critical points* $\mathbf{x}_0$. (a) If the $n \times n$ submatrices of the bordered Hessian are all negative for $n > 1$, then $\mathbf{x}_0$ is at a local minimum.

   (b) If the signs alternate: positive, negative, positive, ... , starting with the 2×2 matrix, then $\mathbf{x}_0$ is at a local maximum.

   (c) If the pattern doesn't satisfy (a) or (b), and the submatrices are not all zero, then $\mathbf{x}_0$ is at a saddle point.

   Note that the sign of the submatrices is opposite of those for the test developed in the previous section.

8. *Extrema on a region.* The method of Lagrange multipliers is only good for locating extrema on a boundary. Don't forget to analyze the critical points inside the region.

SOLUTIONS TO SELECTED EXERCISES

1. Use the method of Lagrange multipliers. We have $\nabla f(x, y, z) = (1, -1, 1)$, and the constraint is $g(x, y) = x^2 + y^2 + z^2 - 2$, so $\lambda \nabla g(x, y, z) = \lambda(2x, 2y, 2z)$. Thus, $\nabla f = \lambda \nabla g$ gives us $1 = \lambda 2x$, $-1 = \lambda 2y$, and $1 = \lambda 2z$. So we have $x = z = -y = 1/2\lambda$. Substitute this into the constraint: $(1/2\lambda)^2 + (-1/2\lambda)^2 + (1/2\lambda)^2 = 3/4\lambda^2 = 2$,

or $\lambda = \pm(1/2)\sqrt{3/2}$. For $\lambda = +(1/2)\sqrt{3/2}$, we have $(x, y, z) = (\sqrt{2/3}, -\sqrt{2/3}, \sqrt{2/3})$, and for $\lambda = -(1/2)\sqrt{3/2}$, we have $(x, y, z) = (-\sqrt{2/3}, \sqrt{2/3}, -\sqrt{2/3})$. These are the two extreme points and the maximum value is $\sqrt{6}$, while the minimum value is $-\sqrt{6}$.

3. We want to find the extrema of $f(x, y) = x$ subject to $x^2 + 2y^2 = 3$. Use the method of Lagrange multipliers. From the constraint, let $g(x, y) = x^2 + 2y^2 - 3$, so $\nabla f = (1, 0)$ and $\nabla g = (2x, 4y)$. We want to simultaneously solve $\nabla f = \lambda \nabla g$ and the constraint equation:

$$1 = \lambda 2x \qquad (1)$$
$$0 = \lambda 4y \qquad (2)$$
$$x^2 + 2y^2 = 3 \qquad (3)$$

From (2), we get $y = 0$. From (1), $x = 1/2\lambda$. Substituting for x and y in (3) gives us $(1/2\lambda)^2 = 3$, so $1/2\lambda = \pm\sqrt{3}$; therefore, $x = \pm\sqrt{3}$. At $(\sqrt{3}, 0)$, $f(x) = \sqrt{3}$ and at $(-\sqrt{3}, 0)$, $f(x) = -\sqrt{3}$. We conclude that the maximum occurs at $(\sqrt{3}, 0)$ and the minimum is at $(-\sqrt{3}, 0)$.

8. 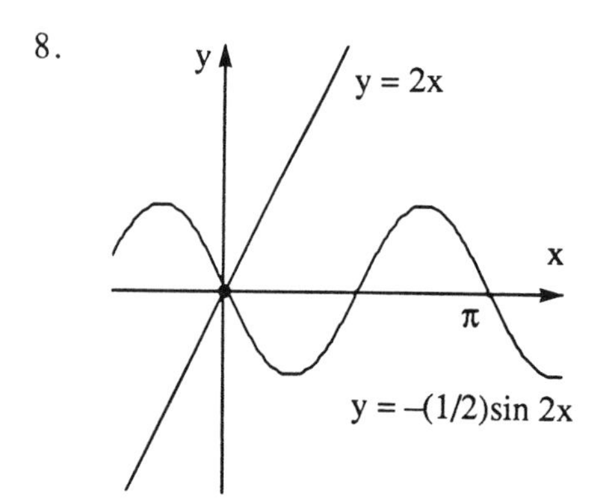

On S , y is restricted to be $\cos x$, so $f(x, y) = x^2 - y^2 = x^2 - \cos^2 x$. Applying one-variable methods, we calculate $f'(x) = 2x + 2\cos x \sin x$. The derivative vanishes when $x = -\cos x \sin x = -(1/2)\sin 2x$. This is a transcendental equation and can be solved by graphical methods (or program an HP-15C to do it for you, if you have one). The graphs of $y = 2x$ and $y = -\sin 2x/2$ only intersect at the origin, so $(0, 0)$ is an extremum. Since $x^2 \geq 0$ and $0 \leq \cos^2 x \leq 1$, we conclude that $(0, 0)$ is a minimum.

13. We want to minimize the surface area of the cylinder subject to the constraint of the volume. That is, we want to minimize $S(r, h) = 2\pi rh + 2\pi r^2$ subject to $\pi r^2 h = 1000\ \text{cm}^3$. Use the method of Lagrange multipliers. From the constraint, we get $g(r, h) = \pi r^2 h - 1000$. We compute the following first partial derivatives: $\partial S/\partial r = 2\pi h + 4\pi r$, $\partial g/\partial r = 2\pi rh$, $\partial S/\partial h = 2\pi r$, $\partial g/\partial h = \pi r^2$. Now, we want to solve the following system of equations:

$$2\pi h + 4\pi r = \lambda 2\pi rh \qquad (1)$$
$$2\pi r = \lambda \pi r^2 \qquad (2)$$
$$\pi r^2 h = 1000 \qquad (3)$$

Factor out πr from (2) to get $2 = \lambda r$ or $\lambda = 2/r$. Substitution into (1) and factoring out 2π gives $h + 2r = (2/r)rh = 2h$, or $h = 2r$. Substitution into (3) gives $\pi r^2 h = \pi r^2 \cdot 2r = 2\pi r^3 = 1000$, so $r = 10/(2\pi)^{1/3}$ and $h = 2r = 20/(2\pi)^{1/3}$. To check that our result satisfies the constraint, we calculate

$$\pi r^2 h = \pi \frac{100}{(2\pi)^{2/3}} \cdot \frac{20}{(2\pi)^{1/3}} = 1000 .$$

So the desired cylinder has height $20/(2\pi)^{1/3}$ cm and base radius $10/(2\pi)^{1/3}$ cm .

16. Use the method of Lagrage multipliers. Let

$$f(x, y, \lambda) = x + 2y \sec\theta + \lambda(xy + y^2 \tan\theta - A) .$$

Then $\partial f/\partial x = 1 + \lambda y$; $\partial f/\partial y = 2\sec\theta + \lambda(x + 2y\tan\theta)$; and $\partial f/\partial\lambda = xy + y^2\tan\theta - A$. From $\partial f/\partial x = 0$, we get $\lambda = -1/y$; whereas from $\partial f/\partial y = 0$, we get

$$\lambda = \frac{-2\sec\theta}{x + 2y\tan\theta} .$$

Hence, $2y = (x + 2y\tan\theta)/(\sec\theta)$, so $2y(1 - \sin\theta) = x\cos\theta$. Thus, $x = 2y(\sec\theta - \tan\theta)$. Substitute this into $\partial f/\partial\lambda = 0$ to get $2y^2(\sec\theta - \tan\theta) + y^2\tan\theta = A$. Then $y^2(2\sec\theta - \tan\theta) = A$, so

$$y^2 = \frac{A}{2\sec\theta - \tan\theta} = \frac{A\cos\theta}{2 - \sin\theta} .$$

23. (a) Use the method of Lagrange multipliers on the auxiliary function $h(x, y, \lambda) = x + y^2 - \lambda(2x^2 + y^2 - 1)$. We compute the following partial derivatives:

$$h_x = 1 - 4x\lambda = 0 \qquad (1)$$

$$h_y = 2y - 2y\lambda = 0 \qquad (2)$$

$$h_\lambda = -(2x^2 + y^2 - 1) = 0 \qquad (3)$$

From (2), we get either $y = 0$ or $\lambda = 1$. For the case where $\lambda = 1$, we get $x = 1/4$ from (1) and then $y = \pm\sqrt{7/8}$ from (3). When $y = 0$, we get $x = \pm 1/\sqrt{2}$ from (3). Thus, the critical points for the constrained function are located at $(1/4, \pm\sqrt{7/8})$ and $(\pm 1/\sqrt{2}, 0)$.

(b) We let the constraint be $g(x, y) = 2x^2 + y^2 - 1$, so $h(x, y) = f(x, y) - \lambda g(x, y)$. Then the bordered Hessian (theorem 9) is

$$|\overline{H}| = \begin{vmatrix} 0 & -\partial g/\partial y & -\partial g/\partial y \\ -\partial g/\partial x & \partial^2 h/\partial x^2 & \partial^2 h/\partial y\partial x \\ -\partial g/\partial y & \partial^2 h/\partial x\partial y & \partial^2 h/\partial y^2 \end{vmatrix} = \begin{vmatrix} 0 & -4x & -2y \\ -4x & -4\lambda & 0 \\ -2y & 0 & 2 - 2\lambda \end{vmatrix} .$$

At $(x, y, \lambda) = (1/4, \sqrt{7/8}, 1)$, we have

$$|\overline{H}| = \begin{vmatrix} 0 & -1 & -\sqrt{7/2} \\ -1 & -4 & 0 \\ -\sqrt{7/2} & 0 & 0 \end{vmatrix} = 14 > 0 ,$$

so $(x, y) = (1/4, \sqrt{7/8})$ is a relative maximum point. At the point $(x, y, \lambda) = (1/4, -\sqrt{7/8}, 1)$, we have

$$|\overline{H}| = \begin{vmatrix} 0 & -1 & \sqrt{7/2} \\ -1 & -4 & 0 \\ \sqrt{7/2} & 0 & 0 \end{vmatrix} = 14 > 0 ,$$

so $(x, y) = (1/4, -\sqrt{7/8})$ is also a relative maximum point. When $(x, y) = (1/\sqrt{2}, 0)$, equation (1) above tells us that $\lambda = -\sqrt{1/8}$, so at the point $(x, y, \lambda) = (1/\sqrt{2}, 0, -\sqrt{1/8})$, we have

$$|\overline{H}| = \begin{vmatrix} 0 & -\sqrt{8} & 0 \\ -\sqrt{8} & \sqrt{2} & 0 \\ 0 & 0 & 2+\sqrt{2} \end{vmatrix} = (2+\sqrt{2})(-8) < 0 .$$

Thus, $(1/\sqrt{2}, 0)$ is a relative minimum. Similarly, when $(x, y) = (-1/\sqrt{2}, 0)$, $\lambda = +\sqrt{1/8}$, so at the point $(x, y, \lambda) = (-1/\sqrt{2}, 0, \sqrt{1/8})$, we have

$$|\overline{H}| = \begin{vmatrix} 0 & \sqrt{8} & 0 \\ \sqrt{8} & -\sqrt{2} & 0 \\ 0 & 0 & 2-\sqrt{2} \end{vmatrix} = (2-\sqrt{2})(-8) < 0 .$$

Thus, $(-1/\sqrt{2}, 0)$ is also a relative minimum. Evaluating the function $f(x, y, z)$ at the critical points tells us that $(1/4, \pm\sqrt{7/8})$ are absolute maxima, $(-1/\sqrt{2}, 0)$ is an absolute minimum, and $(1/\sqrt{2}, 0)$ is only a relative minimum.

4.4 The Implicit Function Theorem

GOALS

1. Be able to determine if an inverse function exists near a point $\mathbf{x}$.
2. If an inverse exists, be able to find a derivative by implicit methods.

STUDY HINTS

1. *Advanced material.* The theorems presented in this section are usually proved in more advanced courses. You should be most concerned with understanding the statements of the theorems.
2. *Notation.* $\mathbf{D}_{\mathbf{x}}F$ is used in this section. It is just another notation for ∇F with respect to $\mathbf{x}$.
3. *Local theorems.* The theorems introduced in this section may not apply if the range or domain is too large.

4. *Special implicit function theorem.* If $\partial F/\partial z \neq 0$, z can be written in terms of $\mathbf{x}$ at a given point $(\mathbf{x}_0, z_0)$, and the derivative of $z = g(\mathbf{x})$ is
$$\mathbf{D}g(\mathbf{x}) = \frac{-\mathbf{D}_\mathbf{x}F(\mathbf{x}, z)}{\frac{\partial F}{\partial z}(\mathbf{x}, z)} .$$
Note that we can differentiate z even though we don't have a formula for z. Don't forget the minus sign in the derivative.

5. *Commonly used formula.* When z is a function of x and y, we get the formula
$$\frac{dy}{dx} = -\frac{\partial z/\partial x}{\partial z/\partial y} .$$
It looks almost like division of fractions, except for the minus sign. Again, don't forget the minus sign.

6. *General implicit function theorem.* In general, $\mathbf{z}$ may be a vector. We form a matrix with the top row being the partials of F_1 with respect to z_j, $j = 1, \dots, m$ (almost like a gradient). Similarly, the other rows consist of the partials of F_k, $k = 1, \dots, m$. If the determinant of this $m \times m$ matrix is non-zero, then $\mathbf{z}$ is a function of $\mathbf{x}$ and a derivative exists.

7. *Jacobian.* This is the *determinant* of the $m \times m$ matrix described in item 6 above. It is denoted $Jf(\mathbf{x}_0)$ or $\frac{\partial(f_1, \dots, f_m)}{\partial(x_1, \dots, x_m)}$.

8. *Inverse function theorem.* As stated in item 6 above, if $Jf(\mathbf{x}_0) \neq 0$, then $\mathbf{z}$ can be written in terms of $\mathbf{x}$. It may not be easy, but it is possible.

9. *Example 3.* This is a typical problem. Study it carefully. Note that when more than one function is given, you will need to solve a system of simultaneous equations to find a partial derivative.

SOLUTIONS TO SELECTED EXERCISES

2. Let $F(x, y, z) = xy + z + 3xz^5 - 4$. Since we want to know if we can solve for z as a function of (x, y), we need to know that $\partial F/\partial z$ does not vanish near the desired point, so $\partial F/\partial z = 1 + 15xz^4$. Near $(1, 0, 1)$, $F_z = 16 \neq 0$, so $F = 0$ is solvable for z as a function of (x, y). Therefore,
$$\frac{\partial z}{\partial x} = \frac{-F_x}{F_z} = -\frac{y + 3z^5}{1 + 15xz^4} \quad \text{and}$$
$$\frac{\partial z}{\partial y} = \frac{-F_y}{F_z} = -\frac{x}{1 + 15xz^4} .$$
At $(x, y) = (1, 0)$, $z = 1$, so $\partial z/\partial x = -3/16$ and $\partial z/\partial y = -1/16$.

7. Let $F_1 = y + x + uv$ and $F_2 = uxy + v$. Then we want

$$\Delta = \begin{vmatrix} \partial F_1/\partial u & \partial F_1/\partial v \\ \partial F_2/\partial u & \partial F_2/\partial v \end{vmatrix} \neq 0 \text{ at } (x, y, u, v) = (0, 0, 0, 0) .$$

The entries of the determinant are $\partial F_1/\partial u = v$, $\partial F_1/\partial v = u$, $\partial F_2/\partial u = xy$, and $\partial F_2/\partial v = 1$. We see that at $(0, 0, 0, 0)$,

$$\Delta = \begin{vmatrix} v & u \\ xy & 1 \end{vmatrix} = \begin{vmatrix} 0 & 0 \\ 0 & 1 \end{vmatrix} = 0 ,$$

so we may not solve for u, v in terms of x, y near $(x, y, u, v) = (0, 0, 0, 0)$. To check directly, the first equation gives us $uv = -(y + x)$, so $v = -(y + x)/u$. Combining this with the second equatio, we get $uxy = (x + y)/u$, or $u^2 = (x + y)/xy$. For (x, y) near $(0, 0)$, either there is no solution for u small, or there are 2 solutions for u.

10. (a) Using the definition of $\partial(x, y)/\partial(r, \theta)$ and computing the partial derivatives, we get

$$\frac{\partial(x, y)}{\partial(r, \theta)} = \begin{vmatrix} \partial x/\partial r & \partial x/\partial\theta \\ \partial y/\partial r & \partial y/\partial\theta \end{vmatrix} = \begin{vmatrix} \cos\theta & -r\sin\theta \\ \sin\theta & r\cos\theta \end{vmatrix} = r .$$

At (r_0, θ_0), $r = r_0$.

(b) By the inverse function theorem, we can form a smooth inverse function $(r(x, y), \theta(x, y))$ as long as $r \neq 0$. As a direct check, solve for r and θ in terms of x and y: $x^2 + y^2 = r$ and $\tan\theta = y/x$ or $\theta = \arctan(y/x)$. Since we have written r and θ in terms of x and y, the result above is confirmed. Note that if $x = 0$ then $\theta = \pi/2$ or $3\pi/2$, depending on the sign of y. If, in addition, $y = 0$, then $r = 0$, and θ can have any value, so we cannot find an inverse, as we did above.

12. Let $F_1 = xy^2 + xzu + yv^2 - 3$ and $F_2 = u^3yz + 2xv - u^2v^2 - 2$. Then $\partial F_1/\partial u = xz$, $\partial F_1/\partial v = 2yv$, $\partial F_2/\partial u = 3u^2yz - 2uv^2$, $\partial F_2/\partial v = 2x - 2u^2v$. At $(x, y, z) = (1, 1, 1)$ and $(u, v) = (1, 1)$,

$$\Delta = \left|\frac{\partial(F_1, F_2)}{\partial(x, y)}\right| = \begin{vmatrix} xz & 2yv \\ 3u^2yz - 2uv^2 & 2x - 2u^2v \end{vmatrix} = \begin{vmatrix} 1 & 2 \\ 1 & 0 \end{vmatrix} = -2 \neq 0 .$$

Since $\Delta \neq 0$, it is possible to solve for u, v in terms of x, y, z near the given point. To compute $\partial v/\partial y$, we use the chain rule:

$$\frac{\partial F_1}{\partial y} = 2xy + xz\frac{\partial u}{\partial y} + v^2 + 2yv\frac{\partial v}{\partial y} = 0 \text{ and}$$

$$\frac{\partial F_2}{\partial y} = 3u^2\frac{\partial u}{\partial y}yz + u^3z + 2x\frac{\partial v}{\partial y} - 2u\frac{\partial u}{\partial y}v^2 - 2v\frac{\partial v}{\partial y}u^2 = 0 .$$

At $(x, y, z) = (1, 1, 1)$ and $(u, v) = (1, 1)$, those equations become $2 + \partial u/\partial y + 1 + 2\,\partial v/\partial y = 0$ and $3\,\partial u/\partial y + 1 + 2\,\partial v/\partial y - 2\,\partial u/\partial y - 2\,\partial v/\partial y = 0$, or $\partial u/\partial y + 2\,\partial v/\partial y = -3$ and $\partial u/\partial y = -1$, so $\partial v/\partial y = -1$.

4.5 Some Applications

GOALS

1. For mechanics problems, be able to find an equilibrium point and determine its stability.
2. Be able to find minimum and maximum distances in geometric problems.
3. For economics problems, be able to explain the significance of λ.

STUDY HINTS

1. *Optional material.* Your instructor may not wish to include this section in the course outline. Use your lectures or ask your instructor to decide which applications, if any, to concentrate on.

2. *Mechanics.* You should know the following facts. An equilibrim point is where $\mathbf{F}(\mathbf{x}_0) = 0$. These are the critical points of a potential V. An equilibrium point is stable if it is a strict local minimum of V. These facts are also true if the potential is constrained.

3. *Geometry.* As in example 7, section 4.2, we can analyze the square of the distance rather than the distance itself.

4. *Economics.* Isoquants are curves showing all possible combinations of capital and labor which produce the *same output.* The Lagrange multiplier λ usually has no special significance. However, in these examples, λ tells you how much more can be produced with one extra unit of capital *or* labor. Note that λ has significance only at the optimal point.

SOLUTIONS TO SELECTED EXERCISES

1. An equilibrium point occurs at a critical point. Compute the partials of V and set them equal to 0:
$$\partial V/\partial x = 6x + 2y + 2 = 0 \qquad (1)$$
$$\partial V/\partial y = 2x + 2y + 1 = 0 \qquad (2)$$
From (1), $2y = -6x - 2$. Substitute into (2) to get $2x + (-6x - 2) + 1 = 0$, so $-4x = 1$ or $x = -1/4$. Then $y = -1/4$. For a *stable* equilibrium, we have to show that $(x, y) = (-1/4, -1/4)$ is a strict local minimum of V: The second partials are $\partial^2V/\partial x^2 = 6$ and $\partial^2V/\partial x\partial y = \partial^2V/\partial y\partial x = \partial^2V/\partial y^2 = 2$. So the discriminant is
$$D = \begin{vmatrix} 6 & 2 \\ 2 & 2 \end{vmatrix} = 12 - 4 = 8 .$$
Since the Hessian is positive definite, $(-1/4, -1/4)$ is a local minimum.

5. We want to minimize $V = mgz + x + y$ subject to the constraint $x^2 + y^2 + z^2 = 1$. Use the method of Lagrange multipliers to get the following system of equations:

$1 = 2x\lambda$ (1)

$1 = 2y\lambda$ (2)

$mg = 2z\lambda$ (3)

$x^2 + y^2 + z^2 = 1$ (4)

From (1) and (2), we get $x = y = 1/2\lambda$. From (3), we get $z = mg/2\lambda = mgx = mgy$. Substitution into (4) gives us $(1/2\lambda)^2 + (1/2\lambda)^2 + (mg/2\lambda)^2 = 1$ or $(1/4\lambda^2)(2 + m^2g^2) = 1$, and $\lambda^2 = (2 + m^2g^2)/4$ or $\lambda = \pm\sqrt{2 + m^2g^2}/2$. Then $x = \pm 1/\sqrt{2 + m^2g^2}$, $y = \pm 1/\sqrt{2 + m^2g^2}$, and $z = \pm mg/\sqrt{2 + m^2g^2}$, and we get two points: $(1/\sqrt{2 + m^2g^2}, 1/\sqrt{2 + m^2g^2}, mg/\sqrt{2 + m^2g^2})$ and $(-1/\sqrt{2 + m^2g^2}, -1/\sqrt{2 + m^2g^2}, -mg/\sqrt{2 + m^2g^2})$. At the first point,

$$V = \frac{(mg)^2}{\sqrt{2 + m^2g^2}} + \frac{2}{\sqrt{2 + m^2g^2}} = \frac{2 + (mg)^2}{\sqrt{2 + m^2g^2}}.$$

At the second point,

$$V = \frac{-(mg)^2}{\sqrt{2 + m^2g^2}} - \frac{2}{\sqrt{2 + m^2g^2}} = -\frac{2 + (mg)^2}{\sqrt{2 + m^2g^2}}.$$

Since V is continuous on the unit sphere, it is clear that $(-1/\sqrt{2 + m^2g^2}, -1/\sqrt{2 + m^2g^2}, -mg/\sqrt{2 + m^2g^2})$ is a stable equilibrium point.

8. With a hyperbola, there is only a minimum distance from a point. With a parabola, there is also only a minimum distance. There can be no maximum distance from these geometric figures because both figures extend to infinity.

11. Let the price of labor be p and let the price of capital be q. We want to optimize Q given the constraint $S = pL + qK = B$. We compute the partials of Q and S: $\partial Q/\partial K = A\alpha K^{\alpha-1}L^{1-\alpha}$, $\partial S/\partial K = q$, $\partial Q/\partial L = A(1-\alpha)L^{-\alpha}K^{\alpha}$, $\partial S/\partial L = p$. Use the method of Lagrange multipliers to get the following system of equations:

$A\alpha K^{\alpha-1}L^{1-\alpha} = \lambda q$ (1)

$A(1-\alpha)K^{\alpha}L^{-\alpha} = \lambda p$ (2)

$pL + qK = B$. (3)

From (1), we get $q = (A/\lambda)\alpha K^{\alpha-1}L^{1-\alpha}$, and from (2), we get $p = (A/\lambda) \times (1-\alpha)K^{\alpha}L^{-\alpha}$. Substitution into (3) gives us $(A/\lambda)(1-\alpha)K^{\alpha}L^{1-\alpha} + (A/\lambda)\alpha K^{\alpha}L^{1-\alpha} = B$ or $(A/\lambda)K^{\alpha}L^{1-\alpha} = B$ or $\lambda = (A/B)K^{\alpha}L^{1-\alpha}$. Substitute for λ in (1) and (2):

$A\alpha K^{\alpha-1}L^{1-\alpha} = (A/B)K^{\alpha}L^{1-\alpha}q$ (4)

$A(1-\alpha)K^{\alpha}L^{\alpha} = (A/B)K^{\alpha}L^{1-\alpha}p$ (5)

From (4), we get $K = \alpha B/q$, and from (5), we get $L = (1-\alpha)B/p$. So the point

$$(K, L) = \left(\frac{\alpha B}{q}, \frac{(1-\alpha)B}{p}\right)$$

optimizes the profit.

4.R Review Exercises for Chapter 4

SOLUTIONS TO SELECTED EXERCISES

1. (b) Compute the partials $\partial z/\partial x = 2x + Cy$ and $\partial z/\partial y = 2y + Cx$. We see that $(0, 0)$ is indeed a critical point. The second partials of z are $\partial^2 z/\partial x^2 = 2$, $\partial^2 z/\partial y^2 = 2$, and $\partial^2 z/\partial x\partial y = C$. By theorem 5, if $4 - C^2 > 0$ we get a relative minimum. If $4 - C^2 = 0$, then $C = \pm 2$, and $z = x^2 + y^2 \pm 2xy = (x \pm y)^2 \geq 0$. In this case, z is equal to 0 if and only if $x = y = 0$. So $(0, 0)$ is a minimum if $4 - C^2 = 0$. If $4 - C^2 < 0$, then we have a saddle point. In summary, $(0, 0)$ is a local minimum if $-2 \leq C \leq 2$; otherwise, $(0, 0)$ is a saddle point.

2. (c) Compute the partials of f and set them equal to 0:

$$f_x = 2x + y^2 = 0 \qquad (1)$$
$$f_y = 2xy + 4y^3 = 0 \qquad (2)$$

From (1), $y^2 = -2x$ (so $x \leq 0$). Substitute into (2) to get $y(-y^2) + 4y^3 = 3y^3 = 0$ or $y = 0$. Thus, $x = 0$ also and $(0, 0)$ is the critical point. The reader should verify that the discriminant D is 0, so this test is inconclusive. However, $f(x, y) = x^2 + 2xy^2 + y^4 - xy^2 = (x + y^2)^2 - xy^2$, so $f(x, y) > 0$ for all $x < 0$, and for x positive, $(x + y^2)^2 < xy^2$ implies $x^2 + y^4 < -xy^2$ but this is impossible, since $x^2 + y^4$ is always positive and $-xy^2$ is always negative. So we can conclude that the minimum value of $f(x, y)$ is 0 at $(0, 0)$.

5. By the method of Lagrange multipliers, we get the following system of equations:

$$y = \lambda$$
$$x = \lambda$$
$$x + y = 1.$$

Substituting into the last equation gives us $2\lambda = 1$, or $\lambda = 1/2$. Thus, $x = 1/2$ and $y = 1/2$. However, in this case, it would have been much easier to solve for x and substitute to get $z = x(1 - x) = x - x^2$. By one-variable calculus methods, $x = 1/2$ is the critical point, and the maximum value of z subject to $x + y = 1$ is $1/4$.

8. (b) By the implicit function theorem,

$$\frac{dy}{dx} = -\frac{\partial F/\partial x}{\partial F/\partial y}.$$

In this case, $F(x, y) = x^3 - \sin y + y^4 - 4 = 0$. So $dy/dx = 3x^2/(-\cos y + 4y^3)$.

10. (c) The rectangular parallelepiped(box) is symmetric, so if x is a coordinate of a point (x, y, z) on the corner of the box, then the dimension corresponding to that coordinate has to be $2x$. We want to maximize $V(x, y, z) = (2x)(2y)(2z) = 8xyz$ subject to $x^2/a^2 + y^2/b^2 + z^2/c^2 = 1$. Use the method of Lagrange multipliers:

$$\partial V/\partial x = 2yz = 8x\lambda/a^2 \qquad (1)$$
$$\partial V/\partial y = 2xz = 8y\lambda/b^2 \qquad (2)$$
$$\partial V/\partial z = 2xy = 8z\lambda/c^2 \qquad (3)$$
$$x^2/a^2 + y^2/b^2 + z^2/c^2 = 1 \qquad (4)$$

Solve (1) for x: $x = 4a^2yz/\lambda$. Substitute into (2): $8(4a^2yz/\lambda)z = 2y\lambda/b^2$, or z^2

$= \lambda^2/16a^2b^2$. Substitute for x in (3): $8(a^2y^2/\lambda)y = 2z\lambda/c^2$, or $y^2 = \lambda^2/16a^2c^2$. Then

$$x^2 = \frac{16a^4}{\lambda^2}y^2z^2 = \frac{16a^4}{\lambda^2}\left(\frac{\lambda^2}{16a^2c^2}\right)\left(\frac{\lambda^2}{16a^2b^2}\right) = \frac{\lambda^2}{16b^2c^2}.$$

Plug these results into (4) to get

$$\frac{\lambda^2}{16a^2b^2c^2} + \frac{\lambda^2}{16a^2b^2c^2} + \frac{\lambda^2}{16a^2b^2c^2} = 1.$$

This simplifies to $3\lambda^2 = 16a^2b^2c^2$, or $\lambda^2 = 16a^2b^2c^2/3$. Substitution for λ^2 gives us

$$x^2y^2z^2 = \frac{a^3}{3} \bullet \frac{b^3}{3} \bullet \frac{c^3}{3} = \frac{a^2b^2c^2}{27},$$

or $xyz = abc/3\sqrt{3}$. Therefore, the maximum volume is

$$V = 8xyz = 8 \bullet \frac{abc}{3\sqrt{3}} = \frac{8abc}{3\sqrt{3}}.$$

14. First, check for extrema on the interior of the circle of radius $\sqrt{2}$. We have $f(x, y) = xy - y + x - 1$, so set its partial derivatives equal to 0. We have

$$\partial f/\partial x = y + 1 \quad \text{and} \quad \partial f/\partial y = x - 1,$$

so the point $(-1, 1)$ is an extremum. Since $(-1)^2 + (1)^2 = 2$, this point is within the constraint, and $f(-1, 1) = -4$. To find other extrema (if any), we use the method of Lagrange multipliers:

$$y + 1 = 2x\lambda \qquad (1)$$
$$x - 1 = 2y\lambda \qquad (2)$$
$$x^2 + y^2 = 2 \qquad (3)$$

From (1) and (2), we get $\lambda = (y + 1)/2x$ and $\lambda = (x - 1)/2y$. Set them equal, cross multiply, and simplify to get

$$y^2 - x^2 + y + x = 0 \qquad (4)$$

From (3), $y^2 = 2 - x^2$. Substitute into (4) to get $2 - x^2 - x^2 + y + x = 0$ or $y = 2x^2 - x - 2$. Substitute back for y in (3): $(2x^2 - x^2 - 2)^2 + x^2 = 2$. This simplifies to $2x^4 - 2x^3 - 3x^2 + 2x + 1 = 0$, and factoring yields $2x^2(x^2 - 1) - 2x(x^2 - 1) - (x^2 - 1) = 0$ or $(2x^2 - 2x + 1)(x^2 - 1) = 0$. The reader should verify that $x = \pm 1$ are the only real solutions. Finally, substitute into (3) and we get $y = \pm 1$. So the four extrema are $(1, 1)$, $(-1, 1)$, $(1, -1)$, and $(-1, -1)$. We compute $f(1, 1) = f(1, -1) = f(-1, -1) = 0$ and $f(-1, 1) = -4$. Thus the maximum points are $(1, 1)$, $(-1, -1)$, and $(1, -1)$, with maximum value of 0. The minimum point is $(-1, 1)$, with the minimum value of -4. [Note: this problem illustrates a rare case where an "interior" extremum coincides with one on the boundary; in practice, one should *always* check for interior extrema.]

17. We use the implicit function theorem. We need to show that the determinant

$$\begin{vmatrix} \partial F/\partial u & \partial F/\partial v \\ \partial G/\partial u & \partial G/\partial v \end{vmatrix}$$

is not 0 near $(x, y, u, v) = (2, -1, 2, 1)$. The determinant is

$$\begin{vmatrix} -3u^2 & 2v \\ -4u & 12u^3 \end{vmatrix} = \begin{vmatrix} -12 & 2 \\ -8 & 12 \end{vmatrix} = -144 + 16 \neq 0 .$$

Since the determinant is not 0, u and v exist as functions of x and y. To compute $\partial u/\partial x$, we implicitly differentiate the given equations with respect to x. Keep in mi that u and v are functions of x and y. We get

$$2x - 3u^2 \frac{\partial u}{\partial x} + 2v \frac{\partial v}{\partial x} = 0$$

$$2y - 4u \frac{\partial u}{\partial x} + 12v^3 \frac{\partial v}{\partial x} = 0$$

To make the calculation simpler, we can plug in $(x, y, u, v) = (2, -1, 2, 1)$. Then

$$4 - 12 \frac{\partial u}{\partial x} + 2 \frac{\partial v}{\partial x} = 0$$

$$-2 - 8 \frac{\partial u}{\partial x} + 12 \frac{\partial v}{\partial x} = 0 .$$

Solve this simple system of two equations by your favorite method. You should get $\partial u/\partial x = 13/32$.

22. (a) Using the given formula, we write $s = f(m, b) = (1 - 1\cdot m - b)^2 + (3 - 2m - b)^2 + (3 - 4m - b)^2 = 19 - (46m + 16b) + ((m + b)^2 + (2m + b)^2 + (4m + b)^2)$. The problem is to find m and b which minimize $f(m, b)$, so we take derivatives:

$$\frac{\partial f}{\partial m} = -46 + 2(m + b) + 4(2m + b) + 8(4m + b)$$

$$= -46 + 42m + 14b,$$

and

$$\frac{\partial f}{\partial b} = -16 + 2(m + b) + 2(2m + b) + 2(4m + b)$$

$$= -16 + 14m + 6b.$$

Next, we set them equal to 0:

$$-46 + 42m + 14b = 0$$

$$-16 + 14m + 6b = 0 .$$

Solving this system of 2 equations, we get $b = 1/2$, $m = 13/14$. The reader is encouraged to verify that this is indeed a minimum point. Therefore, the best-fitting straight line to the points (1, 1), (2, 3), (4, 3) is $y = 13x/14 + 1/2$, as shown in the graph below.

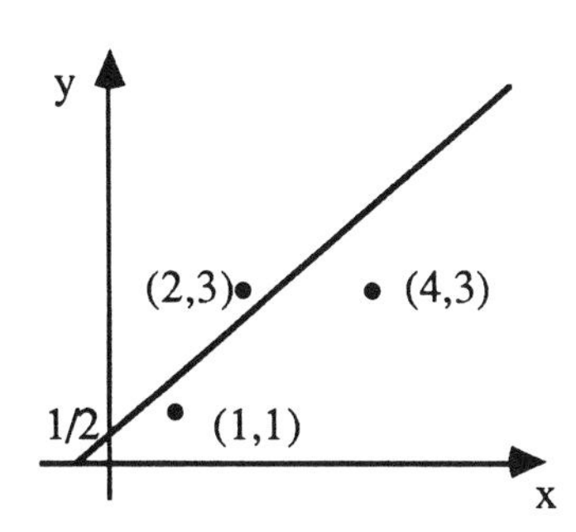

25. If $y = mx + b$ is the best-fitting straight line, we must have, in particular,

$$\frac{\partial}{\partial b}\left(\sum_{i=1}^{n}(y_i - mx_i - b)^2\right) = 0.$$

Performing this differentiation, we get

$$-2\sum_{i=1}^{n}(y_i - mx_i - b) = 0,$$

which implies that the summation has to be 0, or the positive and the negative deviations cancel.

CHAPTER 5
DOUBLE INTEGRALS

5.1 Introduction

GOALS

1. Be able to calculate double integrals over rectangles.

STUDY HINTS

1. *Some notation.* (a) The cartesian product of two intervals in $\mathbf{R}^2$ is a rectangle. If $a \le x \le b$ and $c \le y \le d$, then it is denoted $[a, b] \times [c, d]$.
   (b) Sometimes, dx dy or dy dx is abbreviated dA.

2. *Review.* Before continuing, you should review integration techniques for one variable. It is essential that you remember how to integrate by parts and by substitution.

3. *Geometric interpretation.* If $f(x, y) \ge 0$, then the double integral $\iint_D f(x, y)\, dA$ is the volume under the surface defined by the graph of $z = f(x, y)$. Recall, that with one-variable, $\int f(x)\, dx$ was the area under the curve $y = f(x)$.

4. *Computing a double integral.* Just as with partial derivatives, all but one variable is held constant in each step. We integrate from the inside and work our way to the outside. For example,
$$\int_a^b \int_c^d f(x, y)\, dy\, dx = \int_a^b [F(x, d) - F(x, c)]\, dx ,$$
where F is an antiderivative of f when x is held constant. Now, we compute the integral using one-variable methods.

5. *Cavalieri's principle.* This principle is used in most one-variable calculus courses to derive volume formulas. These formulas are often referred to by the names: disk or slice method.

6. *Riemann sums.* Be aware that in the summation $\sum_{i=0}^{n} f(c_i)(x_{i+1} - x_i)$, c_i can be chosen *anywhere* in $[x_{i+1}, x_i]$. In a Riemann sum, we take the limit as $n \to \infty$, so in most cases, $f(c_i)$ is almost independent of c_i because x_{i+1} and x_i are usually close together.

SOLUTIONS TO SELECTED EXERCISES

1. (b) First hold x constant and integrate with respect to y , then integrate with respect to x :

$$\int_0^{\pi/2}\int_0^1 (y\cos x + 2)\,dy\,dx = \int_0^{\pi/2}\left[\left(\frac{y^2}{2}\cos x + 2y\right)\Big|_{y=0}^{1}\right]dx$$
$$= \int_0^{\pi/2}\left(\frac{1}{2}\cos x + 2\right)dx = \left(\frac{1}{2}\sin x + 2x\right)\Big|_0^{\pi/2} = \frac{1}{2} + \pi\,.$$

2. (b) As you will see later in this chapter, we can easily change the order of integration since the region is a rectangle. Thus, we hold y constant and integrate with respect to x first:

$$\int_0^1\int_0^{\pi/2} (y\cos x + 2)\,dx\,dy = \int_0^1 [(y\sin x + 2x)\Big|_{x=0}^{\pi/2}]\,dy$$
$$= \int_0^1 (y+\pi)\,dy = \left(\frac{y^2}{2} + \pi y\right)\Big|_0^1 = \frac{1}{2} + \pi\,.$$

As expected, we get the same answer regardless of the order of integration.

3. By Cavalieri's principle, the volume of a solid is

$$V = \int_a^b A(x)\,dx\,,$$

where $A(x)$ is the cross-sectional area cut out by a plane. Note that at the same height, a cross-section of the left-hand side is a circle of radius r and a cross-section of the right-hand side is also a circle of radius r . Since $A(x)$ is equal in both cases, the volumes must be equal.

5. With the setup in figure 5.1.12, we slice W vertically by planes to produce triangles R_x of area $A(x)$ in the figure. The base b of the triangle is $b = \sqrt{r^2 - x^2}$ and its height is $h = b\tan\theta = \sqrt{r^2 - x^2}\tan\theta$. Thus, $A(x) = (1/2)bh = (1/2)(r^2 - x^2)\tan\theta$. Hence, the volume is

$$\int_{-r}^{r} A(x)\,dx = \int_{-r}^{r}\frac{1}{2}(r^2 - x^2)\tan\theta\,dx = \frac{1}{2}(\tan\theta)\left(r^2x - \frac{x^3}{3}\right)\Big|_{-r}^{r}$$
$$= \frac{1}{2}\tan\theta\left(2r^3 - \frac{2r^3}{3}\right) = \frac{2r^3}{3}\tan\theta\,.$$

8. Note that when y is in the interval $[-1, 0]$, $|y| = -y$, so

$$\int_R\left(|y|\cos\frac{\pi x}{4}\right)dy\,dx = \int_0^2\int_{-1}^0\left(-y\cos\frac{\pi x}{4}\right)dy\,dx$$
$$= \int_0^2\left[\left(\frac{-y^2}{2}\cos\frac{\pi x}{4}\right)\Big|_{y=-1}^{0}\right]dx$$
$$= \int_0^2\left(\frac{1}{2}\cos\frac{\pi x}{4}\right)dx = \frac{2}{\pi}\sin\frac{\pi x}{4}\Big|_0^2 = \frac{2}{\pi}\,.$$

10. Since $f(x, y) \geq 0$ for all points in R, $\int_R f(x, y)\, dy\, dx$ is the desired volume. It is

$$\int_1^2 \int_0^1 (1 + 2x + 3y)\, dy\, dx = \int_1^2 \left[\left(y + 2xy + \frac{3y^2}{2} \right) \Big|_{y=0}^{1} \right] dx$$

$$= \int_1^2 \left(2x + \frac{5}{2} \right) dx = \left(x^2 + \frac{5x}{2} \right) \Big|_1^2 = \frac{11}{2} .$$

5.2 The Double Integral Over a Rectangle

GOALS

1. Be able to compute a double integral over a rectangular region.

2. Understand Fubini's theorem.

STUDY HINTS

1. *Definition.* The definition of the integral is more important for theoretical rather than computational work. It is defined by the Riemann sum:

$$\iint_R f(x, y)\, dA = \lim_{n \to \infty} \sum_{j, k = 0}^{n-1} f(c_{jk})\, \Delta x\, \Delta y .$$

Although it is not required, it is usually convenient to use a regular partition, i.e., a partition with equal spacings.

2. *Properties.* Many of the properties that hold for single integrals also hold for double integrals. Some of these include the integrability of any continuous or even piecewise continuous function.

3. *Warning.* As in one-variable integration, $[\iint_R f\, dA][\iint_R g\, dA] \neq \iint_R fg\, dA$, in general. In other words, the product of two integrals does not usually equal the integral of the product.

4. *New terminology.* The names of the properties are known as linearity, homogeneity, monotonicity, and additivity. You should know the statements even if you don't know the names.

5. *Fubini's theorem.* This tells you that, for a piecewise continuous function, you can integrate one variable at a time and the order of integration does not matter.

6. *Double integrals and volumes.* Recall that if $f(x, y) \geq 0$, the double integral is simply the volume of the region between the graph of $f(x, y)$ and the xy plane. If $f(x, y) < 0$, then we subtract the corresponding volume between the graph of $f(x, y)$ and the xy plane.

SOLUTIONS TO SELECTED EXERCISES

1. (b) We compute the double integral as an iterated integral:

$$\int_R ye^{xy}\,dA = \int_0^1\int_0^1 ye^{xy}\,dx\,dy = \int_0^1 [e^{xy}\Big|_{x=0}^{1}]\,dy$$

$$= \int_0^1 (e^y - 1)dy = (e^y - y)\Big|_0^1 = e - 2\,.$$

Note that we chose to integrate in x first because integrating with respect to y first would require integration by parts.

2. (b) This is the iterated integral $\int_0^1\left[\int_0^1 (ax + by + c)\,dx\right]dy =$

$$\int_0^1\left[\left(\frac{ax^2}{2} + (by + c)\,x\right)\Big|_{x=0}^{1}\right]dy = \int_0^1\left(\frac{a}{2} + by + c\right)dy = \left[\frac{by^2}{2} + \left(\frac{a}{2} + c\right)y\right]\Big|_0^1$$

$= a/2 + b/2 + c$.

5. We will use the fact that $\int cf(x)\,dx = c\int f(x)\,dx$ for any constant c . If we integrate in x first, then g(y) is held constant in the first step, and so

$$\int_R [f(x)g(y)]\,dx\,dy = \int_c^d\int_a^b f(x)g(y)\,dx\,dy = \int_c^d\left[\int_a^b f(x)g(y)\,dx\right]dy =$$

$$\int_c^d\left[g(y)\int_a^b f(x)\,dx\right]dy\,,\quad\text{where}\quad \int_a^b f(x)\,dx \text{ is a constant in } y\,.$$

Factoring out the constant integral gives us

$$\left[\int_a^b f(x)\,dx\right]\left[\int_c^d g(y)\,dy\right]\,.$$

7. The function $x^2 + y$ is positive over R , so the double integral represents the desired volume.

$$\int_0^1\int_1^2 (x^2 + y)\,dy\,dx = \int_0^1\left[\left(yx^2 + \frac{y^2}{2}\right)\Big|_{y=1}^{2}\right]dx = \int_0^1\left(x^2 + \frac{3}{2}\right)dx$$

$$= \left(\frac{x^3}{3} + \frac{3x}{2}\right)\Big|_0^1 = \frac{1}{3} + \frac{3}{2} = \frac{11}{6}\,.$$

Thus, the volume is 11/6 .

11. In the first version of Fubini's theorem (theorem 3), a continuous function is required; however, theorem 3′ generalizes the result. The second version states that discontinuities are allowed if f is bounded and is composed of a finite union of continuous functions. The problem in this exercise is that the requirements of a finite union iand boundedness of f not met. Even though f is continuous on each subregion, there is an infinite number of subregions. Also, as one approaches the origin, f goes to $\pm\infty$ in a not-too-well-behaved way.

5.3 The Double Integral Over More General Regions

GOALS

1. Be able to compute a double integral over an elementary region on the xy plane.

STUDY HINTS

1. *Region types.* A type 1 region is bounded by the lines $x = a$, $x = b$, and two curves which are functions of x for $a \leq x \leq b$. A type 2 region is bounded by the lines $y = c$, $y = d$, and two curves which are functions of y for $c \leq y \leq d$. Type 3 regions may be classified as both type 1 and type 2 regions. See figure 5.3.3.

2. *Knowing region types.* For purposes of integrating, it is more important to be able to recognize a region type rather than being able to name it. Your primary concern should be learning how to perform double integration.

3. *Simplifying complicated regions.* Most plane regions may be broken up into regions of type 1 and type 2. For example, the region at the left is divided into six pieces, each of which is either type 1 or type 2. In fact, many of the subregions are both type 1 *and* type 2.

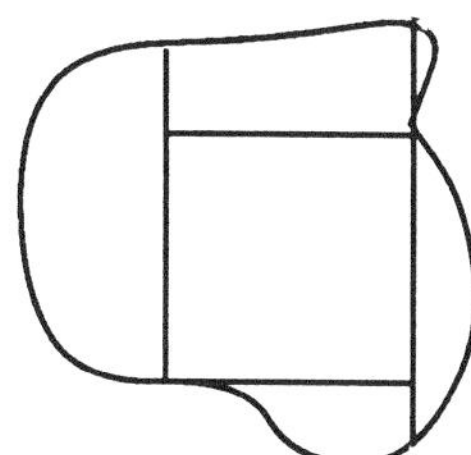

4. *How to integrate.* It is best to use the iterated integral $\int_a^b \int_{\varphi_1(x)}^{\varphi_2(x)} f(x, y)\, dy\, dx$ for type 1 regions. For type 2 regions, it is best to use the iterated integral $\int_c^d \int_{\psi_1(y)}^{\psi_2(y)} f(x, y)\, dx\, dy$.

5. *Choosing integration limits.* When you perform multiple integration, be sure that the limits of integration do not include any previously integrated variables. In particular, no variable should appear in the limits of the outer most integral. For example, the integral of $(x + y)^2$ over the region D: $0 \leq y \leq x^2$, $0 \leq x \leq 1$, should be written as
$$\int_0^1 \int_0^{x^2} (x + y)^2\, dy\, dx \text{, } not \int_0^{x^2} \int_0^1 (x + y)^2\, dx\, dy .$$
Sketching the region often helps in choosing your limits. Also, it helps to read the picture as follows, in this example: "while y ranges from 0 to x^2, x ranges between 0 and 1" or "for each x between 0 and 1, y ranges between 0 and x^2". Draw a picture and think about it.

6. *Definiton of the integral.* As in the last section, this is only important for theoretical purposes. Up to this section, we only know how to integrate over rectangles, so we

cover a region D with a large rectangle and let $f(x, y) = 0$ outside of D.

7. *Double integrals and area.* If $f(x, y) = 1$ on a region D, then $\iint_D f(x, y)\, dA$ is the area of D.

SOLUTIONS TO SELECTED EXERCISES

1. (b)

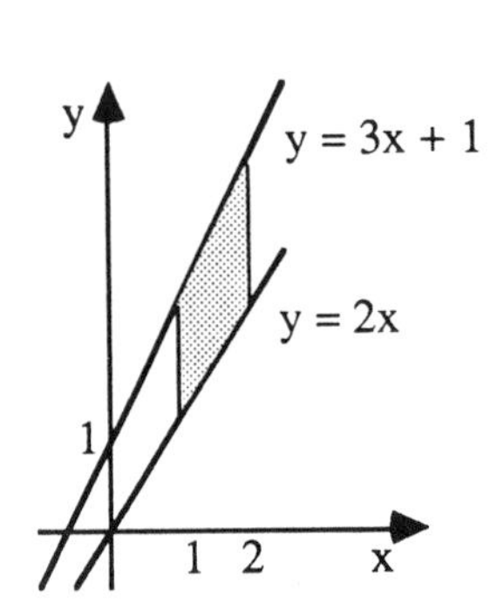

The iterated integral is

$$\int_1^2 \left[\int_{2x}^{3x+1} dy\right] dx = \int_1^2 \left(y\Big|_{y=2x}^{3x+1}\right) dx$$

$$= \int_1^2 (3x + 1 - 2x)\, dx = \int_1^2 (x + 1)\, dx$$

$$= \left(\frac{x^2}{2} + 1\right)\Big|_1^2 = \frac{5}{2}.$$

At each x_0 in $[1, 2]$, the region extends from $2x_0$ to $3x_0 + 1$, so we get the region shown in the sketch. This is a type 1 region because it can described by

$$1 \le x \le 2 \quad \text{and} \quad 2x \le y \le 3x + 1.$$

It is also a type 2 region because it can be described by $2 \le y \le 7$ and $1 \le x \le y/2$ for x in $[2, 4]$, $(y - 1)/3 \le x \le 2$ for x in $[4, 7]$.

2. (b) First, we will sketch the region. At each x_0 in $[-1, 1]$, the region extends from $-2\,|\,x_0\,|$ to $|\,x_0\,|$, so we get the region in the sketch. The integration is most easily done by dividing the region into two parts, so

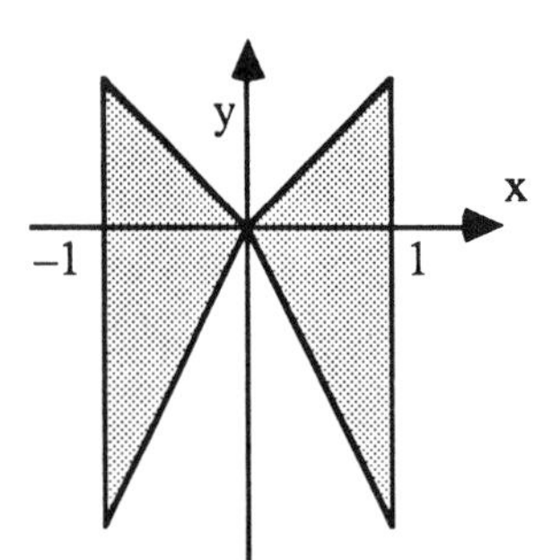

$$\int_{-1}^{1}\int_{-2|x|}^{|x|} e^{x+y}\, dy\, dx$$

$$= \int_{-1}^{0}\int_{2x}^{-x} e^{x+y}\, dy\, dx + \int_{0}^{1}\int_{-2x}^{x} e^{x+y}\, dy\, dx.$$

Note that when x is in $[-1, 0]$, $|\,x\,| = -x$. The first integral of the sum is

$$\int_{-1}^{0} \left(e^{x+y}\Big|_{y=2x}^{-x}\right) dx = \int_{-1}^{0} (1 - e^{3x})dx = \left(x - \frac{e^{3x}}{3}\right)\Big|_{-1}^{0} = \frac{2}{3} - \frac{1}{3e^3}.$$

The second integral of the sum is

$$\int_{0}^{1} \left(e^{x+y}\Big|_{y=-2x}^{x}\right) dx = \int_{0}^{1} (e^{2x} - e^{-x})\, dx = \left(\frac{e^{2x}}{2} + e^{-x}\right)\Big|_{0}^{1} = \frac{e^2}{2} + \frac{1}{e} - \frac{3}{2}.$$

Therefore, the entire integral is $e^2/2 + 1/e - 1/3e^3 - 5/6$. The region is type 1 because it can be described as

$$-1 \le x \le 1 \quad \text{and} \quad -2\,|\,x\,| \le y \le |\,x\,|.$$

It can not be described as a type 2 region without subdividing the region.

(e) The iterated integral is

$$\int_0^1\left[\int_{y^2}^{y}(x^n+y^m)\,dx\right]dy=\int_0^1\left[\left(\frac{x^{n+1}}{n+1}+xy^m\right)\Big|_{x=y^2}^{y}\right]dy$$

$$=\int_0^1\left(\frac{y^{n+1}}{n+1}+y^{m+1}-\frac{y^{2n+2}}{n+1}-y^{m+2}\right)dy$$

$$=\left(\frac{y^{n+2}}{(n+1)(n+2)}+\frac{y^{m+2}}{m+2}-\frac{y^{2n+3}}{(n+1)(2n+3)}-\frac{y^{m+3}}{m+3}\right)\Big|_0^1$$

$$=\frac{1}{(n+1)(n+2)}+\frac{1}{m+2}-\frac{1}{(n+1)(2n+3)}-\frac{1}{m+3}\,.$$

To sketch the region, for every y_0 in $[0, 1]$, x extends from $x = y_0^2$ to $x = y_0$. This is a type 2 region because it can be described as

$$0 \le y \le 1 \quad\text{and}\quad y^2 \le x \le y\,.$$

This is also a type 1 region because it can be described by

$$0 \le x \le 1 \quad\text{and}\quad x \le y \le \sqrt{x}\,.$$

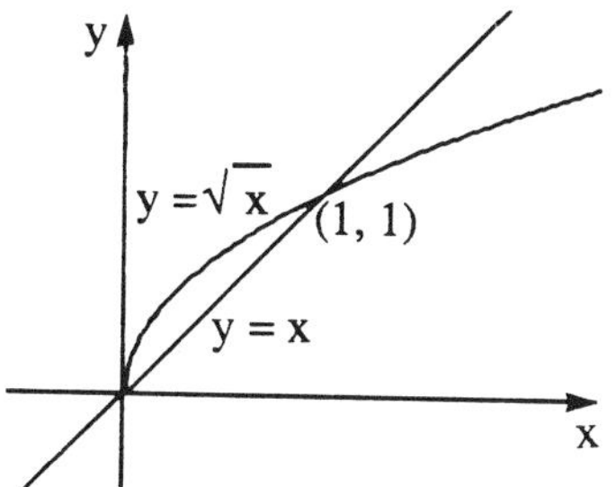

4. The equation of an ellipse with semiaxes a and b is $x^2/a^2 + y^2/b^2 = 1$ or $x^2/b^2 + y^2/a^2 = 1$. In either case, the areas are equal. We will use the first equation, find the area of the region in the first quadrant, and then multiply by 4. The region is described as

$$0 \le x \le a \quad\text{and}\quad 0 \le y \le b\sqrt{1 - x^2/a^2}\,.$$

As a double integral, an area can be computed by $\int_R dA$. In this case, we have

$$\frac{1}{4}A=\int_0^a\int_0^{b\sqrt{1-x^2/a^2}}dy\,dx=\int_0^a\left(y\Big|_{y=0}^{b\sqrt{1-x^2/a^2}}\right)dx=b\int_0^a\sqrt{1-\frac{x^2}{a^2}}\,dx\,.$$

Let $u = x/a$ to get $b\int_0^1 a\sqrt{1-u^2}\,du = ab\int_0^1\sqrt{1-u^2}\,du$. The integral $\int_0^1\sqrt{1-u^2}\,du$ is the area of a quarter of a circle of radius 1, which is $\pi/4$. so $A/4 = ab\pi/4$. Therefore, the area of the ellipse is $A = ab\pi$.

7. The region D is sketched on the next page. It can be described as

$$-\sqrt{3}/2 \le y \le \sqrt{3}/2 \quad\text{and}\quad 0 \le x \le -4y^2 + 3\,,\ \text{so}$$

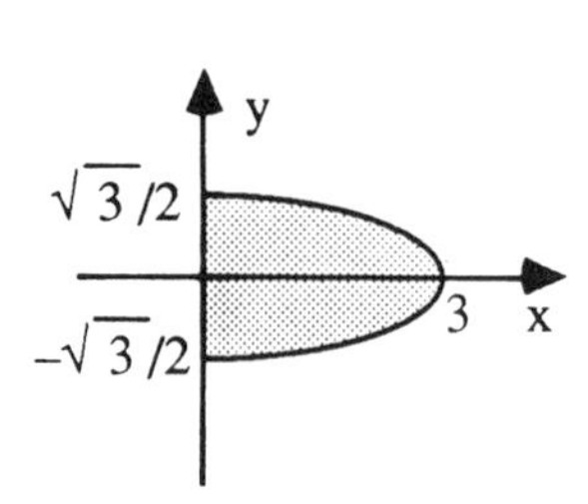

$$\int_D x^3 y\, dx\, dy = \int_{-\sqrt{3}/2}^{\sqrt{3}/2}\int_0^{-4y^2+3} x^3 y\, dx\, dy$$

$$= \int_{-\sqrt{3}/2}^{\sqrt{3}/2}\left(\frac{x^4}{4}y \,\Big|_{x=0}^{-4y^2+3}\right)dy = \int_{-\sqrt{3}/2}^{\sqrt{3}/2} \frac{(4y^2+3)^4 y}{4}\, dy\,.$$

Let $u = -4y^2 + 3$ and we get an integral whose limits of integration are 0 and 0, so the integral is 0.

12.

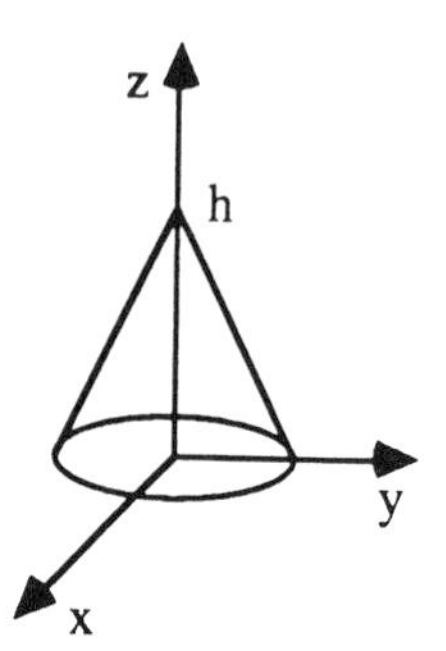

This problem shows the deficiency of integration in Cartesian coordinates. The equation of the cone is:

$$z = h - (h/r)(x^2 + y^2)^{1/2}.$$

The region of integration is the cone's base, which is decribed as

$$-r \le x \le r \quad \text{and} \quad -\sqrt{r^2 - x^2} \le y \le \sqrt{r^2 - x^2}\ .$$

Therefore, the volume of the cone is

$$\int_{-r}^{r}\int_{-\sqrt{r^2-x^2}}^{\sqrt{r^2-x^2}}\left(h - \frac{h}{r}(x^2 + y^2)^{1/2}\right)dy\, dx$$

This is the required answer. (Note: The integration cannot be performed easily at this stage. However, if one peeks ahead in Section 6.3 of the text, one can use cylindrical coordinates (see page 381) and the integral becomes

$$\int_0^{2\pi}\int_0^{r}\left(h - \frac{h}{r}\rho\right)\rho\, d\rho\, d\theta = \frac{\pi r^2}{3}h$$

which is the formula from elementary geometry.)

14. Let D be a type 1 region, then D is described by

$$a \le x \le b \quad \text{and} \quad \varphi_1(x) \le y \le \varphi_2(x)$$

and the integral $\int_D f(x)g(y)\, dA$ becomes

$$\int_a^b \int_{\varphi_1(x)}^{\varphi_2(x)} f(x)g(y)\, dy\, dx\, .$$

Now, we integrate in y. Let G be an antiderivative of g and note that $f(x)$ is a constant in y. We get

$$\int_a^b \left[f(x) \int_{\varphi_1(x)}^{\varphi_2(x)} g(y)\, dy \right] dx = \int_a^b f(x)\, [G(\varphi_2(x)) - G(\varphi_1(x))]\, dx\, .$$

G depends on x, so we can't consider it to be constant and we cannot factor it outside of the integral sign. Thus, in general, $\int_D f(x)g(y)\, dy\, dx$ is not the product of two integrals.

5.4 Changing the Order of Integration

GOALS

1. Be able to evaluate a double intgral by changing the order of integration.

2. Be able to state and understand the mean value theorem for double integrals.

STUDY HINTS

1. *Rationale for changing order.* Recall that Fubini's theorem allows you to change the order of integration. Sometimes, changing the order of integration simplifies a problem. Try doing example 1 without changing the order of integration and compare the efficiency of both methods. (You will probably need to use trigonometric substitution.) At other times, a double integral can only be computed if the order is changed.

2. *Beginning.* It is useful to sketch the region of integration from the given limits before choosing new limits.

3. *Mean value theorem for integrals.* If two conditions hold: (i) f is continuous and (ii) D is an elementary region, then the conclusion is

$$\iint_D f(x, y)\, dA = f(x_0, y_0) \cdot \text{area}(D)$$

for some point (x_0, y_0) in D.

4. *Ingredient in the proof of the mean value theorem.* If m is the minimum value of $f(x, y)$ on D and M is the maximum, then

$$m \cdot \text{area}(D) \le \iint_D f(x, y)\, dA \le M \cdot \text{area}(D)\, .$$

This allows you estimate the value of a double integral.

SOLUTIONS TO SELECTED EXERCISES

1. (b) First, we recall that $\cos^2\theta = (1 + \cos 2\theta)/2$ and compute the integral as written:

$$\int_0^{\pi/2}\int_0^{\cos\theta} \cos\theta \, dr \, d\theta = \int_0^{\pi/2} \left(r\cos\theta \Big|_{r=0}^{\cos\theta}\right) d\theta = \int_0^{\pi/2} \cos^2\theta \, d\theta$$

$$= \int_0^{\pi/2} \left(\frac{1 + \cos 2\theta}{2}\right) d\theta = \left(\frac{\theta}{2} + \frac{\sin 2\theta}{4}\right)\Big|_0^{\pi/2} = \frac{\pi}{4} .$$

From the graph, we see that if we choose an r_0, then θ extends from 0 to $\cos^{-1}(r_0)$. Thus, the region can also be described as

$0 \le r \le 1$ and $0 \le \theta \le \cos^{-1}(r)$.

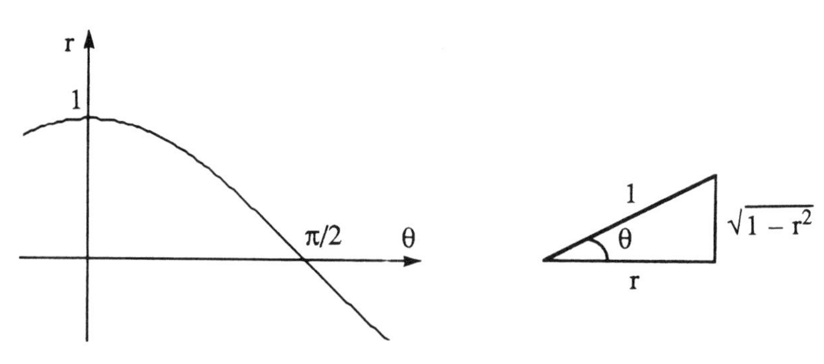

Therefore, changing the order of integration gives us

$$\int_0^1\int_0^{\cos^{-1}(r)} \cos\theta \, d\theta \, dr = \int_0^1 \left(\sin\theta \Big|_{\theta=0}^{\cos^{-1}(r)}\right) dr = \int_0^1 \sin(\cos^{-1}(r)) \, dr .$$

From the triangle, we see that $\sin(\cos^{-1}(r)) = \sqrt{1 - r^2}$. We recognize the integral as the area of a quarter of a circle of radius 1, so

$$\int_0^1 \sqrt{1 - r^2} \, dr = \frac{\pi}{4} .$$

2. (c)

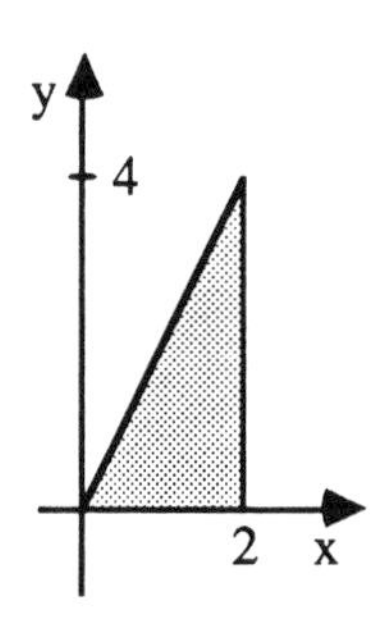

The region of integration is sketched here. There is no function whose derivative is $\exp(x^2)$, so we try changing the order of integration. The region can be expressed as a type 1 region:

$0 \le x \le 2$ and $0 \le y \le 2x$.

So the integral becomes

$$\int_0^2\int_0^{2x} \exp(x^2) \, dy \, dx = \int_0^2 \left[y \exp(x^2)\Big|_{y=0}^{2x}\right] dx .$$

Now let $u = x^2$ and we get

$$\int_0^2 2x \exp(x^2) \, dx = \int_0^4 e^u \, du = e^u \Big|_0^4 = e^4 - 1 .$$

3. This formula is an application of equation (6). The region D is $[-\pi, \pi] \times [-\pi, \pi]$, so A(D) is the area of D, which is $4\pi^2$. The function $f(x, y) = e^{\sin(x + y)}$ is largest when $\sin(x + y)$ is largest, i. e., when $x + y = \pi/2 \pm 2n\pi$, where n is an integer. Similarly, f(x, y) is smallest when $x + y = -\pi/2 \pm 2n\pi$. Over the region D, we have $-1 \le \sin(x + y) \le 1$, so $m = 1/e$ and $M = e^1$. Therefore, substituting everything into formula (6) on page 340 of the text gives us

$$\frac{1}{e} \le \frac{1}{4\pi^2} \int_D f(x, y)\, dA \le e .$$

6. The area of the triangle is 1/2. The function $f(x, y) = 1/(y - x + 3)$ is smallest when $y - x + 3$ is largest on D. Notice that $y \le x$ in D, so $y - x + 3$ is largest when $y = x$, and so $m = 1/3$. Similarly, we see that M occurs at (1, 0) where f(1, 0) = 1/2. Now, formula (6) gives us

$$\frac{1}{3} \le 2\int_D f(x, y)\, dA \le \frac{1}{2} \quad \text{or} \quad \frac{1}{6} \le \int_D f(x, y)\, dA \le \frac{1}{4} .$$

10.

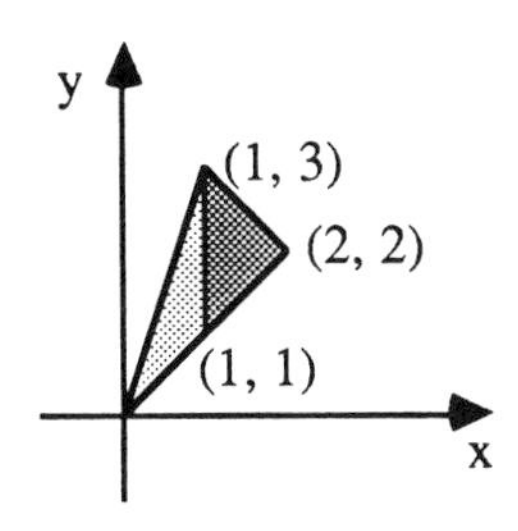

We need to divide the triangle into two parts: one part for x in [0, 1] and the other part for x in [1, 2]. When x is in [0, 1], we have $x \le y \le 3x$, so the integral over this region is

$$\int_0^1 \int_x^{3x} e^{x-y}\, dy\, dx = \int_0^1 \left(-e^{x-y}\Big|_{y=x}^{3x}\right) dx$$

$$= \int_0^1 (-e^{-2x} + 1)\, dx = \left(\frac{e^{-2x}}{2} + x\right)\Big|_0^1$$

$$= \frac{1}{2e^2} + \frac{1}{2} .$$

For the second region, we have $x \le y \le 4 - x$, so the integral over that region is

$$\int_1^2 \int_x^{4-x} e^{x-y}\, dy\, dx = \int_1^2 \left(-e^{x-y}\Big|_{y=x}^{4-x}\right) dx = \int_1^2 (-e^{2x-4} + 1)\, dx$$

$$= \left(\frac{-e^{2x-4}}{2} + x\right)\Big|_1^2 = \frac{1}{2e^2} + \frac{1}{2} .$$

Adding the two integrals together, we get $1 + 1/e^2$.

13.

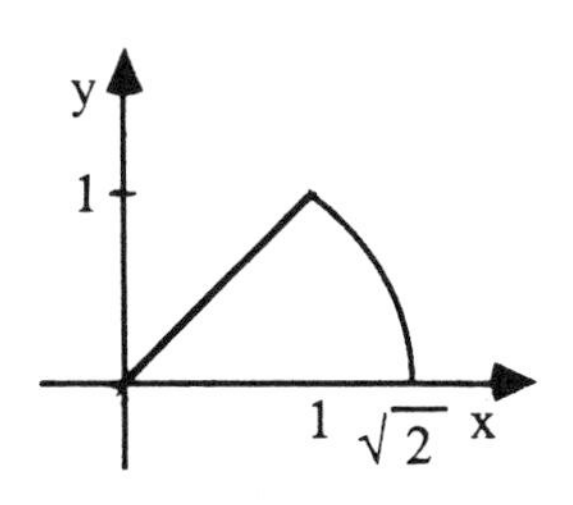

First, we sketch the region. For all y_0 in [0, 1], x goes from y_0 to $\sqrt{2 - y_0^2}$. To describe the region as a type 2 region, we have to subdivide the region at $x = 1$. When x is in [0, 1], we have $0 \le y \le x$. When x is in $[1, \sqrt{2}\,]$, we have $0 \le y \le \sqrt{2 - x^2}$. Thus, we have

$$\int_0^1 \left[\int_y^{\sqrt{2-y^2}} f(x, y)\, dx\right] dy = \int_0^1 \left[\int_0^x f(x, y)\, dy\right] dx + \int_1^{\sqrt{2}} \left[\int_0^{\sqrt{2-x^2}} f(x, y)\, dy\right] dx .$$

In some cases, the left-hand side will be easier to work with.

15. This proof requires using the chain rule. Let

$$G(x, u) = \int_a^x \int_c^d f(u, y, z)\, dz\, dy.$$

We want to find dG/dx (the "total differential", if you will) when $u = x$. By the chain rule,

$$\frac{dG}{dx} = \left.\frac{\partial G}{\partial x}\right|_{u=x} + \left.\frac{\partial G}{\partial u}\right|_{u=x}.$$

$\frac{\partial G}{\partial x}$ is simple: it is equal to $\frac{\partial}{\partial x}\int_a^x \left[\int_c^d f(u, y, z)\, dz\right] dy$. From the fundamental theorem of (one-variable) calculus, we simply take the "integrand" (the dz integral here) and replace y by x (because we are integrating with respect to y). At $u = x$,

$$\left.\frac{\partial G}{\partial x}\right|_{u=x} = \int_c^d f(x, x, z)\, dz.$$

$\frac{\partial G}{\partial u}$ is a little trickier. First,

$$\frac{\partial G}{\partial u} = \frac{\partial}{\partial u}\int_a^x \int_c^d f(u, y, z)\, dz\, dy$$

$$= \int_a^x \int_c^d \frac{\partial}{\partial u} f(u, y, z)\, dz\, dy,$$

assuming that f is a nice function such that the order of integration and differentiation can be interchanged. Now we want to evaulate everything at $u = x$, so we simply replace u by x, and get

$$\left.\frac{\partial G}{\partial u}\right|_{u=x} = \int_a^x \int_c^d \frac{\partial}{\partial x} f(x, y, z)\, dz\, dy.$$

Combine the two results and you've got it.

5.5 Some Technical Integration Theorems

GOALS

1. Be able to understand the proofs of the theorems introduced in chapter 5.

2. Be able to use Riemann sums and Cauchy sequences to do proofs.

STUDY HINTS

1. *How to study this section.* This section has the same level of complexity as section 2.7 and should be studied the same way you studied section 2.7. Ask your instructor if you really have to know it.

2. *Main concepts used in proofs.* Many of the proofs in this section use the concept of the Riemann sum, the triangle inequality, and the ε-δ concept, particularly with the notion of uniform continuity.

3. *Continuity.* In section 2.2, a defintion was given for continuity. That kind of continuity is known as pointwise continuity. In this section, we introduce uniform continuity, a stronger form of continuity. Uniform continuity differs in that δ is chosen independent of any points $\mathbf{x}$ in the domain.

4. *Cauchy sequence.* A Cauchy sequence is a sequence in a point set whose terms s_m and s_n differ by less than ε if n and m are large enough. A basic fact is that every Cauchy sequence converges to a limit. For a closed set, this limit is also in the set.

SOLUTIONS TO SELECTED EXERCISES

1. Suppose $a \neq b$ and $a > b$, then $a - b < \varepsilon$ for all ε. Choose $\varepsilon = (a - b)/2$, then $b + \varepsilon = (a - b)/2 + b = (a + b)/2 < (a + a)/2 = a$, or $a - b > \varepsilon$, which is a contradiction. Hence $a = b$.

5. 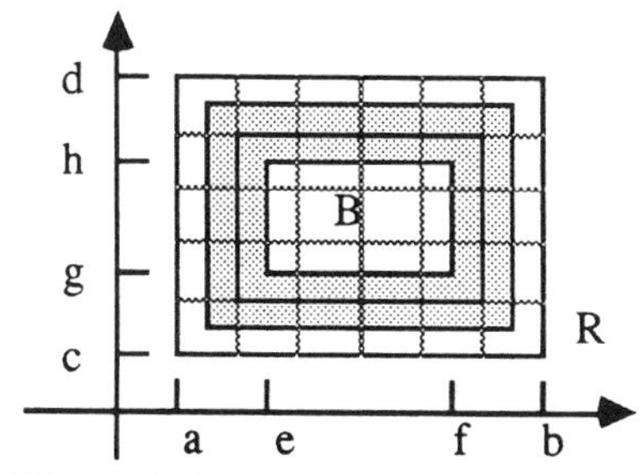

Let $R = [a, b] \times [c, d]$ and $B = [e, f] \times [g, h]$. Partition R into $n \times n$ pieces of rectangles r_n of size $((b - a)/n) \times ((d - c)/n)$, as shown. Obviously, the area of $B \leq b_n$, since b_n is just the area of the union of those r_n that have nonempty intersections with B. Let's call this union A. On the other hand, suppose (x, y) is a point in A. Then (x, y) has to be at least within one rectangle from B, that is, if (x, y) is not in B, it must be somewhere inside a "frame" about B (the shaded area). This gives us $b_n \leq$ area of B + area of frame = area of B + $2(h - g)(b - a)/n + 2(f - e)(d - c)/n + 4((b - a)/n)((d - c)/n)$, or $b_n \leq$ area of B + $(2/n)[(h - g)(b - a) + (f - e)(d - c)] + (4/n^2)[(b - a)(d - c)]$. As $n \to \infty$, $1/n$ and $1/n^2$ both approach 0, so $b_n \leq$ area of B. From this and the earilier result, we conclude that b_n = area of B as $n \to \infty$.

5.R Review Exercises for Chapter 5

SOLUTIONS TO SELECTED EXERCISES

1. (b) Evaluate this as an iterated integral:

$$\int_0^1 \left[\int_{\sqrt{x}}^1 (x + y)^2 \, dy \right] dx = \int_0^1 \left[\frac{(x + y)^3}{3} \Bigg|_{y = \sqrt{x}}^{1} \right] dx$$

$$= \int_0^1 \left[\frac{(x + 1)^3}{3} - \frac{(x + \sqrt{x})^3}{3} \right] dx = \frac{1}{3} \int_0^1 (-3x^{5/2} - x^{3/2} + 3x + 1) dx$$

$$= \frac{1}{3}\left(\frac{-6x^{7/2}}{7} - \frac{2x^{5/2}}{5} + \frac{3x^2}{2} + x\right)\Big|_0^1 = \frac{29}{70} .$$

2. (b)

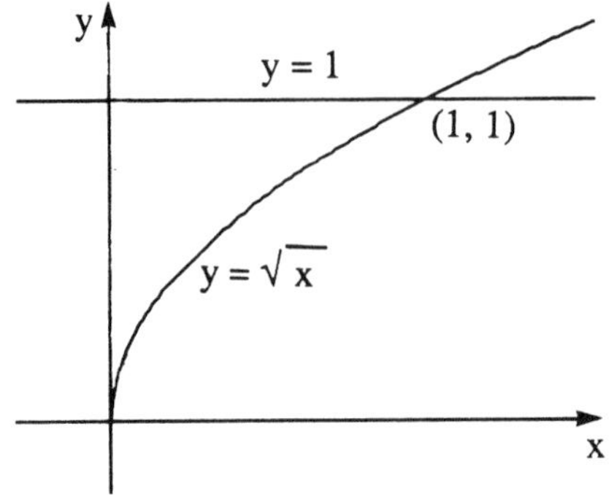

The region is sketched here. As a type 2 region, it can be described as $0 \le y \le 1$ and $0 \le x \le y^2$, so we get

$$\int_0^1 \int_0^{y^2} (x + y)^2 \, dx \, dy$$

$$= \int_0^1 \left[\frac{(x+y)^3}{3} \Big|_{x=0}^{y^2} \right] dy$$

$$= \frac{1}{3}\int_0^1 [(y^2 + y)^3 - y^3] dy$$

$$= \frac{1}{3}\int_0^1 (y^6 + 3y^5 + 3y^4) dy = \frac{1}{3}\left(\frac{y^7}{7} + \frac{y^6}{2} + \frac{3y^5}{5}\right)\Big|_0^1 = \frac{29}{70} .$$

5.

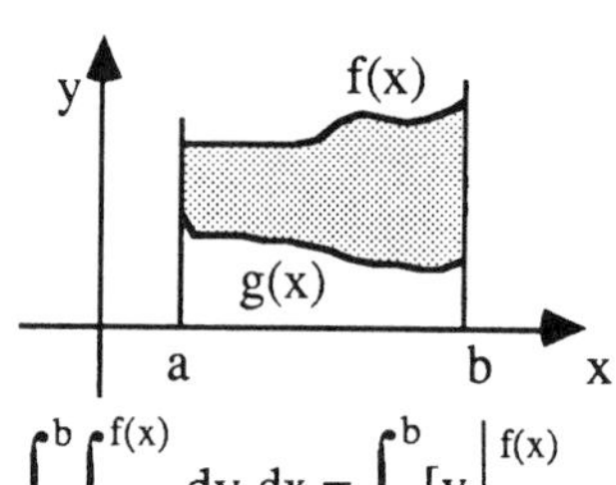

From one-variable calculus, we expect to get

$$\int_a^b [f(x) - g(x)] dx \text{ if } f \ge g \text{ for } x \in [a, b] .$$

The region D can be described as $a \le x \le b$ and $g(x) \le y \le f(x)$ because it is a type 1 region. Therefore, the area of D is

$$\int_a^b \int_{g(x)}^{f(x)} dy \, dx = \int_a^b [y \Big|_{y = g(x)}^{f(x)}] \, dx = \int_a^b [f(x) - g(x)] \, dx .$$

7. (b) As a type 1 region, the unit circle can be described by

$$-1 \le x \le 1 \quad \text{and} \quad -\sqrt{1 - x^2} \le y \le \sqrt{1 - x^2} .$$

Thus, we have

$$\int_{-1}^1 \int_{-\sqrt{1-x^2}}^{\sqrt{1-x^2}} x^2 y^2 \, dy \, dx = \int_{-1}^1 \left[\frac{x^2 y^3}{3} \Big|_{y=-(1-x^2)^{1/2}}^{(1-x^2)^{1/2}} \right] dx$$

$$= \frac{2}{3}\int_{-1}^1 x^2 (1 - x^2)^{3/2} \, dx .$$

Use the trigonometric substitution: $\sqrt{1 - x^2} = \cos\theta$, $x = \sin\theta$, and $dx = \cos\theta \, d\theta$, so

$$\int x^2 (1 - x^2)^{3/2} \, dx = \int \sin^2\theta \cos^3\theta \, (\cos\theta \, d\theta) = \int (\sin^2\theta - 2\sin^4\theta + \sin^6\theta) \, d\theta ,$$

which was obtained by using the identity $\sin^2\theta + \cos^2\theta = 1$ and expanding. Now, use the identities: $\sin^2\theta = (1 - \cos 2\theta)/2$ and $\cos^2\theta = (1 + \cos 2\theta)/2$. We get

$$\int\left[\frac{1-\cos 2\theta}{2}-\frac{1-2\cos 2\theta}{2}-\frac{\cos^2 2\theta}{2}+\frac{1-3\cos 2\theta+3\cos^2 2\theta-\cos^3 2\theta}{8}\right]d\theta$$

$$=\int\left(\frac{1}{16}-\frac{1}{16}\cos 4\theta+\frac{\cos 2\theta\sin^2 2\theta}{8}\right)d\theta=\frac{\theta}{16}-\frac{\sin 4\theta}{64}-\frac{\sin^3 2\theta}{48}+C\,.$$

Now, we use the double angle identities and the following which we get from the triangle:
$\theta=\sin^{-1}x$, $\sin\theta=x$, and $\cos\theta=\sqrt{1-x^2}$.
Substituting and evaluating at ± 1 , we get

$$\left[\frac{\sin^{-1}x}{16}-\frac{4x\sqrt{1-x^2}(1-x^2-x^2)}{64}-\frac{8x^3(1-x^2)\sqrt{1-x^2}}{48}\right]\Bigg|_{-1}^{1}=\frac{\pi}{16}\,.$$

Therefore, the original integral becomes $(2/3)(\pi/16)=\pi/24$.

12.

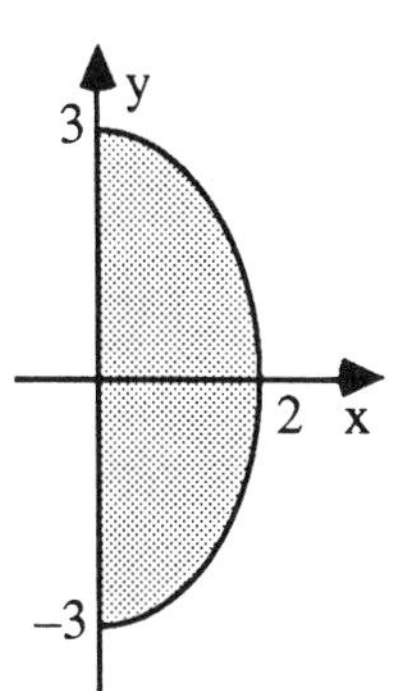

As the integral is written, the region is described as $0\le x\le 2$ and $-3\sqrt{4-x^2}/2\le y\le 3\sqrt{4-x^2}/2$, which is a type 1 region. When we rearrange $y=3\sqrt{4-x^2}/2$, we get $y^2/9+x^2/4=1$, which we recognize as an ellipse, so our sketch is as shown. We evaluate as follows:

$$\int_0^2\int_{-3\sqrt{4-x^2}/2}^{3\sqrt{4-x^2}/2}\left(\frac{5}{\sqrt{2+x}}+y^3\right)dy\,dx$$

$$=\int_0^2\left[\left(\frac{5y}{\sqrt{2+x}}+\frac{y^4}{4}\right)\Bigg|_{y=-(3/2)(4-x^2)^{1/2}}^{(3/2)(4-x^2)^{1/2}}\right]dx$$

$$=\int_0^2\left[15\sqrt{2-x}+\frac{81}{64}(16-8x^2+x^4)\right]dx=\left[-10(2-x)^{3/2}+\frac{81}{64}\left(16x-\frac{8x^3}{3}+\frac{x^5}{5}\right)\right]\Bigg|_0^2$$

$$=20\sqrt{2}+324/15\,.$$

16.

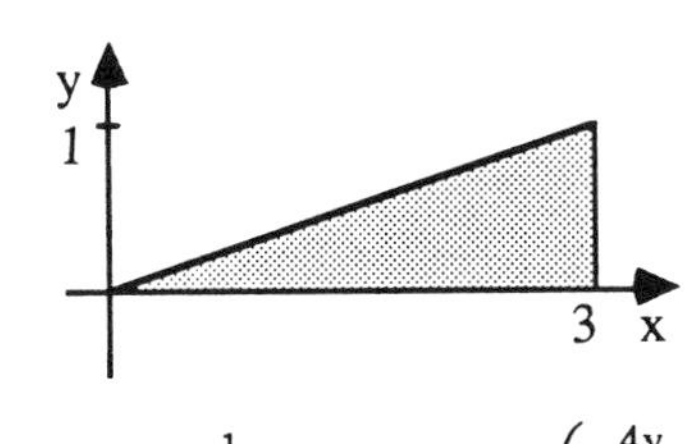

As written, the region described by $0\le y\le 1$ and $0\le x\le 3y$ is a type 2 region. The region is sketched as shown. We evaluate the double integral as follows:

$$\int_0^1\int_0^{3y}e^{x+y}\,dx\,dy=\int_0^1\left(e^{x+y}\Big|_{x=0}^{3y}\right)dy$$

$$=\int_0^1(e^{4y}-e^y)dy=\left(\frac{e^{4y}}{4}-e^y\right)\Bigg|_0^1=\frac{e^4}{4}-e+\frac{3}{4}\,.$$

19. The region is sketched here. Note that it consists of the following two regions:

$0\le x\le 1$ and $0\le y\le -x^2+x$

$1\le x\le 2$ and $-x^2+x\le y\le 0$.

We evaluate as a sum of integrals:

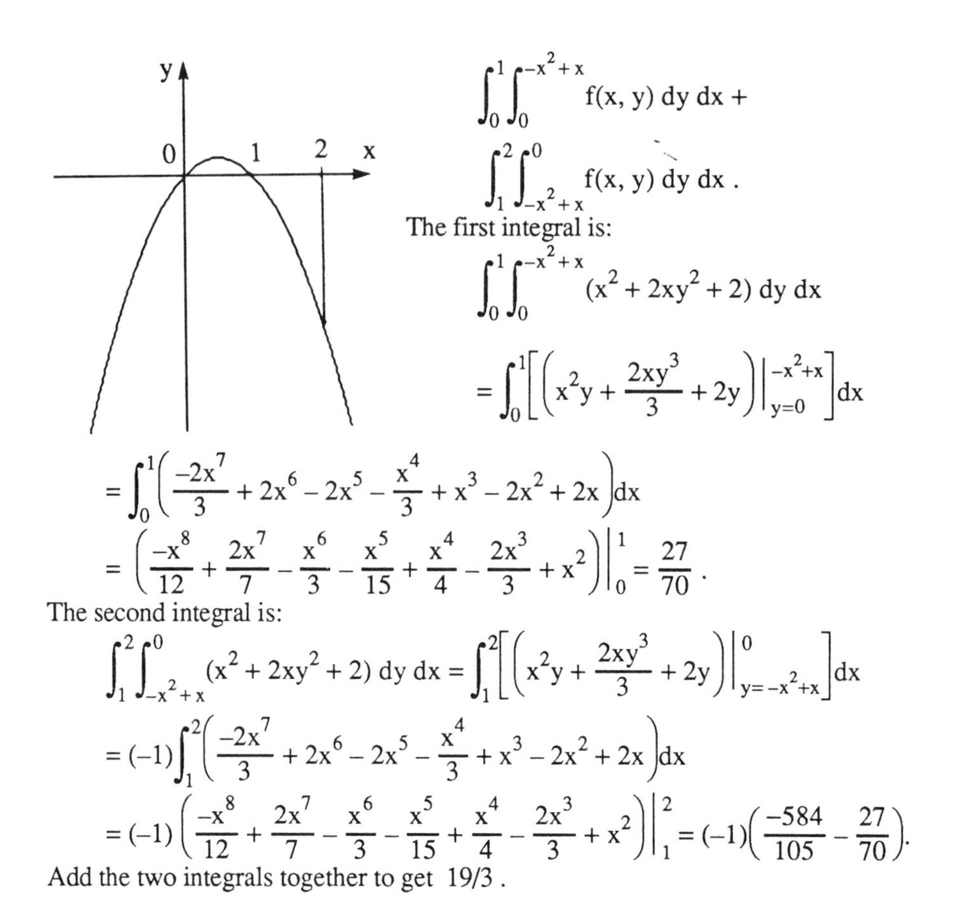

$$\int_0^1 \int_0^{-x^2+x} f(x, y)\, dy\, dx + \int_1^2 \int_{-x^2+x}^0 f(x, y)\, dy\, dx .$$

The first integral is:

$$\int_0^1 \int_0^{-x^2+x} (x^2 + 2xy^2 + 2)\, dy\, dx$$

$$= \int_0^1 \left[\left(x^2 y + \frac{2xy^3}{3} + 2y \right) \Big|_{y=0}^{-x^2+x} \right] dx$$

$$= \int_0^1 \left(\frac{-2x^7}{3} + 2x^6 - 2x^5 - \frac{x^4}{3} + x^3 - 2x^2 + 2x \right) dx$$

$$= \left(\frac{-x^8}{12} + \frac{2x^7}{7} - \frac{x^6}{3} - \frac{x^5}{15} + \frac{x^4}{4} - \frac{2x^3}{3} + x^2 \right) \Big|_0^1 = \frac{27}{70} .$$

The second integral is:

$$\int_1^2 \int_{-x^2+x}^0 (x^2 + 2xy^2 + 2)\, dy\, dx = \int_1^2 \left[\left(x^2 y + \frac{2xy^3}{3} + 2y \right) \Big|_{y=-x^2+x}^{0} \right] dx$$

$$= (-1) \int_1^2 \left(\frac{-2x^7}{3} + 2x^6 - 2x^5 - \frac{x^4}{3} + x^3 - 2x^2 + 2x \right) dx$$

$$= (-1) \left(\frac{-x^8}{12} + \frac{2x^7}{7} - \frac{x^6}{3} - \frac{x^5}{15} + \frac{x^4}{4} - \frac{2x^3}{3} + x^2 \right) \Big|_1^2 = (-1)\left(\frac{-584}{105} - \frac{27}{70} \right).$$

Add the two integrals together to get $19/3$.

23. The function $f(x, y) = x^2 + y^2 + 1$ has the value $r^2 + 1$ where r is the distance from the origin. So on D , $1 \le f(x, y) \le 5$. The area of D is 4π, so by the mean value theorem, we have the desired result.

CHAPTER 6
THE TRIPLE INTEGRAL, CHANGE OF VARIABLES, AND APPLICATIONS

6.1 The Triple Integral

GOALS

1. Be able to compute a triple integral over general regions in space.

2. Be able to change the order of integration for computing triple integrals.

STUDY HINTS

1. *Notation.* We use dV for the differential term $dx\,dy\,dz$ since it represents a volume.

2. *Properties.* Many of the properties of the triple integral are the same as those of the double integral. The triple integral may be considered an iterated integral and we integrate from inside to outside. Fubini's theorem still holds and we can still integrate all piecewise continuous functions.

3. *Region types.* There are two varieties of type I regions, as well as two kinds of type II and type III regions. In many cases, four of the six surfaces are "flat". The region types depend upon the kind of "non-flat" surfaces. Type I "non-flat" surfaces are functions of x and y. Type II "non-flat" surfaces are functions of y and z; type III "non-flat" surfaces are functions of x and z.

4. *Balls.* Example 1 demonstrates that the unit ball is a type I region. Written as
$$-1 \le z \le 1, -\sqrt{1-z^2} \le y \le \sqrt{1-z^2}, \text{ and } -\sqrt{1-z^2-y^2} \le x \le \sqrt{1-z^2-y^2},$$
the ball is a type II region. Similarly, the type III ball is described by
$$-1 \le x \le 1, -\sqrt{1-x^2} \le z \le \sqrt{1-x^2}, \text{ and } -\sqrt{1-x^2-z^2} \le x \le \sqrt{1-x^2-z^2}.$$
The solution of example 1 uses a type 1 description for D. Using a type 2 description would generate three more descriptions of W.

5. *How to integrate.* As with double integrals, knowing how to set up the limits by recognizing the type of region is more important than knowing the type number. Drawing a picture of the region is helpful for finding the limits of integration, although in many instances it is easier said than done.

6. *Triple integrals and volumes.* If $f(x, y, z) = 1$, then $\int_W f\,dV$ is the volume of W.

7. *Integration trick.* A time-saving device is shown in example 1. If we realize that an integral is the area of all or part of a circle, then we don't have to go through the process of finding an antiderivative.

8. *Factoring out integrals.* A triple integral can be easier to compute if you use the following fact: If the limits of integration of the innermost integral do not involve a certain variable and the limits of integration for that particular variable are constant, then that variable may be integrated separately and multiplied by the remaining integral. For example,

$$\int_0^1\int_x^{2x}\int_2^3 f(x)g(y)h(z)\,dz\,dy\,dx = \left(\int_2^3 h(z)\,dz\right)\left(\int_0^1\int_x^{2x} f(x)g(y)\,dy\,dx\right)$$

because the variable z does not appear in any of the limits of integration and its limits of integration are the constants 1 and 0. We cannot integration x separately because it appears in the limits of integration for y.

SOLUTIONS TO SELECTED EXERCISES

2. It is easier to integrate in x before integrating in y. We want to evaluate

$$\int_W e^{-xy}y\,dV = \int_0^1\int_0^1\int_0^1 e^{-xy}y\,dx\,dy\,dz = \int_0^1\int_0^1\left[\left(\frac{e^{-xy}}{-y}\right)y\,\Big|_{x=0}^{1}\right]dy\,dz$$

$$= \int_0^1\int_0^1 (1-e^{-y})\,dy\,dz = \int_0^1\left[(y+e^{-y})\Big|_{y=0}^{1}\right]dz = \frac{1}{e}\int_0^1 dz = \frac{1}{e}\,.$$

Note that you would have had to use integration by parts if you had integrated in y before integrating in x.

5. The plane $x + y = 1$ can be rewritten as $x = 1 - y$, so

$$\int_W x^2\cos z\,dV = \int_0^\pi\int_0^\pi\int_0^{1-y} x^2\cos z\,dx\,dy\,dz = \int_0^\pi\int_0^\pi\left(\frac{x^3}{3}\cos z\,\Big|_{x=0}^{1-y}\right)dy\,dz$$

$$= \frac{1}{3}\int_0^\pi\int_0^\pi (1-y)^3\cos z\,dy\,dz = \frac{1}{3}\int_0^\pi\left[\frac{-(1-y)^4}{4}\cos z\,\Big|_{y=0}^{\pi}\right]dz$$

$$= \frac{1-(1-\pi)^4}{12}\int_0^\pi \cos z\,dz = \frac{1-(1-\pi)^4}{12}(\sin z)\Big|_0^\pi = 0\,.$$

Alternatively, since z does not appear in any of the limits of integration, we can "factor" out the integral in z.

$$\int_W x^2\cos z\,dV = \left(\int_0^\pi\int_0^{1-y} x^2\,dx\,dy\right)\left(\int_0^\pi \cos z\,dz\right) = 0 \text{ since } \int_0^\pi \cos z\,dz = 0\,.$$

7. Evaluate as an iterated integral:

$$\int_0^1\int_0^{2x}\int_{x^2+y^2}^{x+y} dz\,dy\,dx = \int_0^1\int_0^{2x}\left[z\,\Big|_{z=x^2+y^2}^{x+y}\right]dy\,dx = \int_0^1\int_0^{2x}(x+y-x^2-y^2)\,dy\,dx =$$

$$\int_0^1\left[\left((x-x^2)y+\frac{y^2}{2}-\frac{y^3}{3}\right)\Big|_{y=0}^{2x}\right]dx = \int_0^1\left(4x^2-\frac{14x^3}{3}\right)dx = \left(\frac{4x^3}{3}-\frac{7x^4}{6}\right)\Big|_0^1 = \frac{1}{6}\,.$$

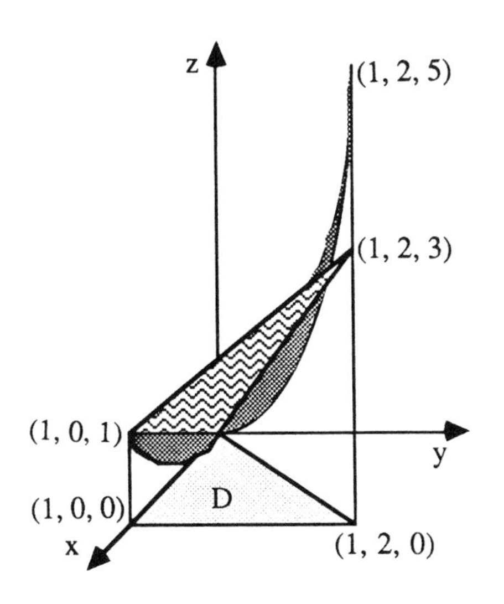

To sketch the region, begin with the interval [0, 1] for x. For each x_0 in the interval, sketch the area D which extends from $y = 0$ to $y = 2x_0$. Then over each point (x_0, y_0) of D, sketch the surfaces $z = x + y$ (shaded with wavy lines) and $z = x^2 + y^2$ (dark shading), which are a plane and a paraboloid. Notice that over part of D, the paraboloid lies above the plane, and over the other part of D, the paraboloid lies below the plane.

10. The surface $z = x^2 + y^2$ is a paraboloid opening upward. The surface $z = 10 - x^2 - 2y^2$ is a paraboloid opening downward. When we equate the two surfaces, we see that they intersect where $x^2 + y^2 = 10 - x^2 - 2y^2$ or $2x^2 + 3y^2 = 10$, which is an ellipse. The ellipse can be described by

$$-\sqrt{5} \le x \le \sqrt{5} \quad \text{and} \quad -\sqrt{10 - 2x^2/3} \le y \le \sqrt{10 - 2x^2/3}\,.$$

As a triple integral, the desired volume is

$$\int_{-\sqrt{5}}^{\sqrt{5}} \int_{-\sqrt{10-2x^2/3}}^{\sqrt{10 - 2x^2/3}} \int_{x^2+y^2}^{10-x^2-2y^2} dz\,dy\,dx = \int_{-\sqrt{5}}^{\sqrt{5}} \int_{-\sqrt{10-2x^2/3}}^{\sqrt{10 - 2x^2/3}} (10 - 2x^2 - 3y^2)\,dy\,dx =$$

$$\int_{-\sqrt{5}}^{\sqrt{5}} \left[(10y - 2x^2y - y^3) \Bigg|_{y = -(10-2x^2)^{1/2}/3}^{(10-2x^2)^{1/2}/3} \right] dx =$$

$$\frac{4}{3\sqrt{3}} \int_{-\sqrt{5}}^{\sqrt{5}} \left[10\sqrt{2}\sqrt{5 - x^2} - 2\sqrt{2}\, x^2\sqrt{5 - x^2} \right] dx\,.$$

From the integral table, we have $\int \sqrt{5 - x^2}\,dx = (x/2)\sqrt{5 - x^2} + (5/2)\sin^{-1}(x/\sqrt{5}) + C$ and $\int x^2\sqrt{5 - x^2}\,dx = (x/8)(2x^2 - 5)\sqrt{5 - x^2} + (25/8)\sin^{-1}(x/\sqrt{5}) + C$. Therefore, we get

$$V = \frac{4}{3\sqrt{3}} \left[10\sqrt{2} \left(\frac{x}{2}\sqrt{5 - x^2} + \frac{5}{2}\sin^{-1}\!\left(\frac{x}{\sqrt{5}}\right)\right) - \right.$$

$$\left. 2\sqrt{2} \left(\frac{x}{8}(2x^2 - 5)\sqrt{5 - x^2} + \frac{25}{8}\sin^{-1}\!\left(\frac{x}{\sqrt{5}}\right)\right)\right] \Bigg|_{-\sqrt{5}}^{\sqrt{5}} = \frac{25\pi\sqrt{2}}{\sqrt{3}}\,.$$

13. The region W is that between the region between the shaded area in the xy plane and the part of the cylinder of radius 2 lying above the shaded area. (See next page.) From our sketch, we see that the shaded area is a type 2 region described by

$0 \le y \le 2$ and $6 - 2y \le x \le 2 - y$.

Finally, z extends from the xy plane, $z = 0$, to the cylinder, $z = \sqrt{4 - y^2}$. Therefore, $\int_W f(x, y, z)\, dV$ is

$$\int_0^2 \int_{2-y}^{6-2y} \int_0^{\sqrt{4-y^2}} z\, dz\, dx\, dy = \int_0^2 \int_{2-y}^{6-2y} \left(\frac{z^2}{2} \Big|_{z=0}^{\sqrt{4-y^2}} \right) dx\, dy$$

$$= \frac{1}{2} \int_0^2 \int_{2-y}^{6-2y} (4 - y^2)\, dx\, dy = \frac{1}{2} \int_0^2 \left[(4 - y^2)x \Big|_{x=2-y}^{6-2y} \right] dy = \frac{1}{2} \int_0^2 (16 - 4y^2 - 4y + y^3)\, dy$$

$$= \frac{1}{2} \left(16y - \frac{4y^3}{3} - 2y^2 + \frac{y^4}{4} \right) \Big|_0^2 = \frac{26}{3}.$$

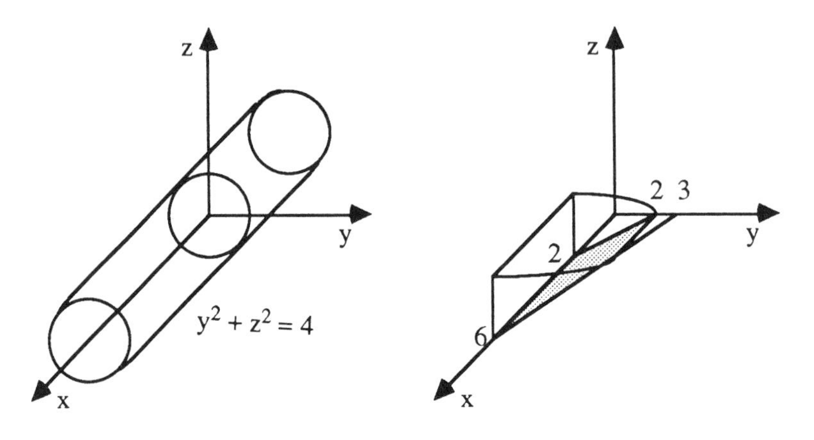

14.

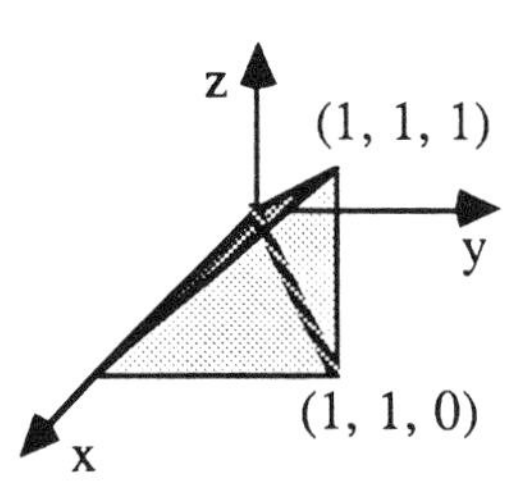

Starting with the interval $0 \le x \le 1$, sketch the area D which extends from $y = 0$ to $y = x_0$ for each x_0 in $[0, 1]$. Then over each point (x_0, y_0) in D, extend the region W from $z = 0$ to $z = y_0$. We shall discuss how to find an equivalent integral with the form

$$\int_W f(x, y, z)\, dx\, dy\, dz.$$

First, note that z goes from 0 to 1. Now, for each z_0, y goes from z_0 to 1. This gives us a triangular area in the yz plane. Finally, for each (y_0, z_0) in the triangular area, x goes from y_0 to 1. Thus, we describe W as follows:

$0 \le z \le 1$ and $z \le y \le 1$ and $y \le x \le 1$.

Note that the surface is bounded by the planes $z = 0$, $x = 1$, $y = z$, and $x = y$. A second way of describing the region is by letting z go from 0 to 1 first. Then x extends from z to 1. And finally y extends from the planes $y = z$ to $y = x$, so W can also be described by

$0 \le z \le 1$ and $z \le x \le 1$ and $z \le y \le x$.

The following integrals are equivalent :

$$\int_0^1\int_0^x\int_0^y f(x, y, z)\, dz\, dy\, dx = \int_0^1\int_y^1\int_0^y f(x, y, z)\, dz\, dx\, dy =$$

$$\int_0^1\int_z^1\int_y^1 f(x, y, z)\, dx\, dy\, dz = \int_0^1\int_0^y\int_y^1 f(x, y, z)\, dx\, dz\, dy =$$

$$\int_0^1\int_0^x\int_z^x f(x, y, z)\, dy\, dz\, dx = \int_0^1\int_z^1\int_z^x f(x, y, z)\, dy\, dx\, dz .$$

As a quick check, we can see if each region has the same volume by letting $f(x, y, z) = 1$. In this case, the volume is $1/6$.

15. Suppose W is a type I region, then

$$\int_a^b\int_{\varphi_1(x)}^{\varphi_2(x)}\int_{-\gamma(x, y)}^{\gamma(x, y)} f(x, y, z)\, dz\, dy\, dx =$$

$$\int_a^b\int_{\varphi_1(x)}^{\varphi_2(x)}\int_0^{\gamma(x, y)} f(x, y, z)\, dz\, dy\, dx + \int_a^b\int_{\varphi_1(x)}^{\varphi_2(x)}\int_{-\gamma(x, y)}^{0} f(x, y, z)\, dz\, dy\, dx .$$

Now, let $z = -u$, so $dz = -du$, and the second half of the sum becomes

$$\int_a^b\int_{\varphi_1(x)}^{\varphi_2(x)}\int_{-\gamma(x, y)}^{0} f(x, y, z)\, dz\, dy\, dx = \int_a^b\int_{\varphi_1(x)}^{\varphi_2(x)}\int_{-\gamma(x, y)}^{0} [-f(x, y, -u)]\, du\, dy\, dx =$$

$$\int_a^b\int_{\varphi_1(x)}^{\varphi_2(x)}\int_0^{-\gamma(x, y)} f(x, y, -u)\, du\, dy\, dx = \int_a^b\int_{\varphi_1(x)}^{\varphi_2(x)}\int_0^{\gamma(x, y)} f(x, y, -z)\, dz\, dy\, dx .$$

We used the fact that $\gamma(x, y) = -\gamma(x, y)$ from the symmetry property and a property of wrong-way integrals. Finally, we changed the dummy variable u back to z . Therefore, the original integral was

$$\int_a^b\int_{\varphi_1(x)}^{\varphi_2(x)}\int_{-\gamma(x, y)}^{\gamma(x, y)} f(x, y, z)\, dz\, dy\, dx =$$

$$\int_a^b\int_{\varphi_1(x)}^{\varphi_2(x)}\int_0^{\gamma(x, y)} [f(x, y, z) + f(x, y, -z)]\, dz\, dy\, dx .$$

Since $f(x, y, z) = -f(x, y, -z)$, the last integrand is 0 . Hence, the entire triple integral is 0 .

18. (a) The volume can be computed as a triple integral with $f(x, y, z) = 1$:

$$\int_0^1\int_0^1\int_0^{xy} dz\, dy\, dx = \int_0^1\int_0^1 \left(z \Big|_{z=0}^{xy} \right) dy\, dx = \int_0^1\int_0^1 xy\, dy\, dx$$

$$= \int_0^1 \left(\frac{xy^2}{2} \Big|_{y=0}^{1} \right) dx = \int_0^1 \frac{x}{2}\, dx = \frac{x^2}{4} \Big|_0^1 = \frac{1}{4} .$$

(d) The integral of z over W is

$$\int_0^1\int_0^1\int_0^{xy} z\, dz\, dy\, dx = \int_0^1\int_0^1 \left(\frac{z^2}{2} \Big|_{z=0}^{xy} \right) dy\, dx = \int_0^1\int_0^1 \frac{x^2y^2}{2}\, dy\, dx$$

$$= \int_0^1 \left(\frac{x^2y^3}{6} \bigg|_{y=0}^{1} \right) dx = \int_0^1 \frac{x^2}{6}\, dx = \frac{x^3}{18} \bigg|_0^1 = \frac{1}{18} .$$

6.2 The Geometry of Maps from $\mathbf{R}^2$ to $\mathbf{R}^2$

GOALS

1. Be able to determine whether a map is one-to-one.
2. Be able to graph a region D which is the image of a region D^* under a mapping.

STUDY HINTS

1. *Notation.* The equality $D = T(D^*)$ expresses the fact that takes a point in D^* and maps it to a point in D, and *all* such points of D are obtained this way. T is called a mapping or a transformation. Thinking of T as a function, this is similar to $y = f(x)$ for $f: \mathbf{R} \to \mathbf{R}$.

2. *One-to-one.* If two distinct points always get mapped to distinct points, i.e., distinct points never get mapped to a single point, then the function is said to be one-to-one. A domain may need to be chosen carefully for this property to hold. (See example 3.) For integration, this is a nice property; otherwise, the double integral over D may not equal the double integral over D^*.

3. *Onto.* If every point of a range D can be mapped by some path in the domain D^*, then the mapping is onto. Onto does not imply one-to-one because two points of D^* may get mapped to one point of D. By the same token, one-to-one does not imply onto either.

4. *Finding* D *from* D^*. In many cases, the range D can be determined by simply mapping the boundary of D^*. Then decide whether the map takes D^* to points inside or outside the boundary of D. For example, let D be the unit disk. If
$$T(x, y) = \left(\frac{1}{x^2 + y^2}, \frac{1}{x^2 + y^2} \right),$$
then T maps the unit circle to the unit circle, and T maps a point such as $(1/2, 1/2)$ to $(2, 2)$, a point outside the circle. So T maps the inside of the circle to the outside (and isn't defined at $(0, 0)$.

5. *Important exercise.* The result of exercise 6 is used in the examples of section 6.3. It states that if $T(\mathbf{x}) = A\mathbf{x}$, then T is one-to-one if and only if $\det A \neq 0$. The result is true for all $n \times n$ determinants.

SOLUTIONS TO SELECTED EXERCISES

2. The transformation is given by the following equation:

$$T(x^*, y^*) = A\mathbf{x} = \begin{bmatrix} 1/\sqrt{2} & -1/\sqrt{2} \\ 1/\sqrt{2} & 1/\sqrt{2} \end{bmatrix}\begin{bmatrix} x^* \\ y^* \end{bmatrix}.$$

We compute $\det A = 1 \neq 0$, so by theorem 1, the linear mapping T maps vertices to vertices. The origin of D^* gets mapped to $T(0, 0) = (0, 0)$ in D. Similarly, the points $(1, 0)$, $(1, 1)$, and $(0, 1)$ of D^* get mapped to $(1/\sqrt{2}, 1/\sqrt{2})$, $(0, \sqrt{2})$, and $(-1/\sqrt{2}, 1/\sqrt{2})$, respectively, in D. It is easy to show that the length of each side of the transformation is 1 and the dot product can be used to show that the angles are $\pi/2$. Thus, T rotates the unit square by 45 degrees.

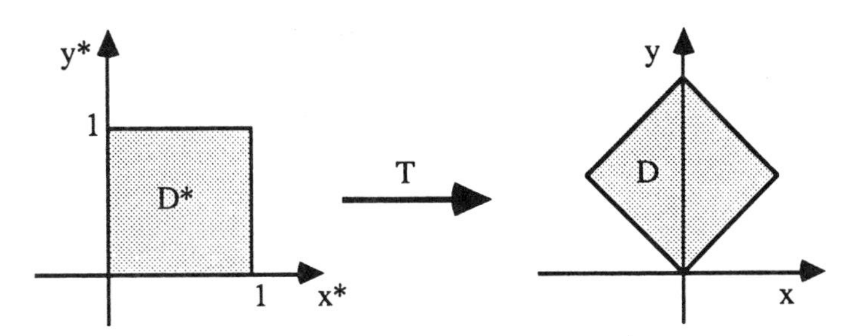

4. The vertices $(0, 0)$, $(1, 0)$, $(1, 1)$, and $(0, 1)$ of D should be mapped from the vertices $(0, 0)$, $(8/5, 4/5)$, $(12/5, 16/5)$, and $(4/5, 12/5)$ of D^*. We want to find a matrix A so that

$$T(x^*, y^*) = A\mathbf{x} = \begin{bmatrix} a & b \\ c & d \end{bmatrix}\begin{bmatrix} x^* \\ y^* \end{bmatrix}.$$

When $(8/5, 4/5)$ gets mapped to $(1, 0)$, we get $8a/5 + 4b/5 = 1$ and $8c/5 + 4d/5 = 0$ so $d = -2c$. When $(4/5, 12/5)$ gets mapped to $(0, 1)$, we get $4a/5 + 12b/5 = 0$ and $4c/5 + 12d/5 = -4c = 1$, or $a = -3b$ and $c = -1/4$, $d = 1/2$. Substituting $a = -3b$ back into $8a/5 + 4b/5 = 1$ gives us $-4b = 1$ or $b = -1/4$, and so $a = 3/4$. Therefore,

$$T(x^*, y^*) = \left(\frac{3}{4}x^* - \frac{1}{4}y^*, \frac{-1}{4}x^* + \frac{1}{2}y^*\right).$$

7. Since sines and cosines appear, we will try to use the identity $\sin^2 t + \cos^2 t = 1$ to eliminate the parameters and obtain a recognizable form: $x^2 + y^2 = \rho^2 \sin^2\varphi\,(\cos^2\theta + \sin^2\theta) = \rho^2 \sin^2\varphi$ and $x^2 + y^2 + z^2 = \rho^2(\sin^2\varphi + \cos^2\varphi) = \rho^2$. We recognize that $x^2 + y^2 + z^2 = \rho^2$ is a sphere of radius ρ; since $0 \leq \rho \leq 1$, D is the unit ball. T is not one-to-one. For examples, $(0, \pi, \pi/2)$ and $(0, \pi/6, \pi/5)$ both map to the origin. Also, $(1, 0, \pi/2)$ and $(1, 2\pi, \pi/2)$ both map to $(1, 0, 0)$.

Since $\rho = 0$ always gets mapped to the origin, we want to eliminate that point. Also, since $\theta = 0$ and $\theta = 2\pi$ give the same mapping, we want to eliminate either one of those points. Hence, T can be made one-to-one by using the following intervals: $\varphi \in$

$[0, \pi]$, $\theta \in (0, 2\pi]$, and $\rho \in (0, 1]$.

10. Write $\mathbf{q} = \mathbf{p} + \lambda\mathbf{v} + \mu\mathbf{w} = (q_1, q_2) = (p_1, p_2) + \lambda(v_1, v_2) + \mu(w_1, w_2)$. Then we have
$$\begin{bmatrix} a_{11} & a_{12} \\ a_{21} & a_{22} \end{bmatrix}\begin{bmatrix} q_1 \\ q_2 \end{bmatrix} = \begin{bmatrix} a_{11}q_1 + a_{12}q_2 \\ a_{21}q_1 + a_{22}q_2 \end{bmatrix}.$$
The first component on the right-hand side can be written as $(a_{11}p_1 + a_{12}p_2) + \lambda(a_{11}v_1 + a_{12}v_2) + \mu(a_{11}w_1 + a_{12}w_2)$. Similarly, we can rewrite the second component. Thus, the vector on the right-hand side can be expressed as $(a_{11}p_1 + a_{12}p_2, a_{21}p_1 + a_{22}p_2) + \lambda(a_{11}v_1 + a_{12}v_2, a_{21}v_1 + a_{22}v_2) + \mu(a_{11}w_1 + a_{12}w_2, a_{21}w_1 + a_{22}w_2)$, which has the form $\mathbf{r} + \lambda\mathbf{s} + \mu\mathbf{t}$. Thus, the transformation does not alter the form of the vectors which describe each point, and it maps parallelograms to parallelograms Recall that as λ , μ vary, we get a parallelogram.

6.3 The Change of Variables Theorem

GOALS

1. Given a transformation T , be able to compute its Jacobian.

2. Be able to use the Jacobian to change variables in double and triple integrals.

3. Be able to state and use the change of variables formula for polar, cylindrical, and spherical coordinates.

STUDY HINTS

1. *Review.* You should review the definitions of polar, cylindrical, and spherical coordinates from section 1.4.

2. *Jacobian.* The Jacobian is a determinant, not a matrix. In this section, we introduce the notation

$\dfrac{\partial(x, y)}{\partial(u, v)}$ and $\dfrac{\partial(x, y, z)}{\partial(u, v, w)}$, which are the determinants $\begin{vmatrix} \partial x/\partial u & \partial x/\partial v \\ \partial y/\partial u & \partial y/\partial v \end{vmatrix}$ and

$\begin{vmatrix} \partial x/\partial u & \partial x/\partial v & \partial x/\partial w \\ \partial y/\partial u & \partial y/\partial v & \partial y/\partial w \\ \partial z/\partial u & \partial z/\partial v & \partial z/\partial w \end{vmatrix}$, respectively.

3. *Changing variables.* As stated, the change of variables theorem requires that T be one-to-one. However, the theorem still holds if T is not one-to-one on the boundary of D*, a situation that occurs in a number of examples. A similar theorem holds for triple integrals.

4. *Useful change of variables.* If $x^2 + y^2$ occurs in the integrand, try using polar or cylindrical coordinates, which have the Jacobian r. If $x^2 + y^2 + z^2$ occurs in the integrand, try using spherical coordinates, which has the Jacobian $\rho^2 \sin \varphi$. These two Jacobians should be memorized.

5. *Limits of integration.* When you set up the limits of integration, remember that φ is measured from the "north pole," *not* from the "equator." Also, recall that θ is measured in a counterclockwise direction.

SOLUTIONS TO SELECTED EXERCISES

2. We know that $x = u + v$ and $y = u - v$. Adding the two equations gives us $(x + y)/2 = u$ and subtraction gives $(x - y)/2 = v$. By mapping the boundaries of the triangle in the xy plane, we get another triangle in the uv plane as shown.

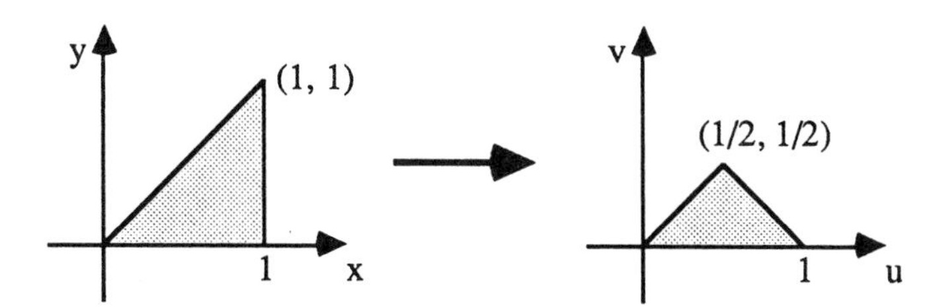

The Jacobian is the absolute value of

$$\left|\frac{\partial(x, y)}{\partial(u, v)}\right| = \begin{vmatrix} \partial x/\partial u & \partial x/\partial v \\ \partial y/\partial u & \partial y/\partial v \end{vmatrix} = \begin{vmatrix} 1 & 1 \\ 1 & -1 \end{vmatrix} = -2,$$

so the Jacobian is 2. In this case, we integrate $x + y = 2u$ as follows:

$$\int_D (x + y)dx\,dy = \int_0^{1/2}\int_v^{1-v} (2u)(2)du\,dv = 4\int_0^{1/2}\int_v^{1-v} u\,du\,dv$$

$$= 4\int_0^{1/2}\left(\frac{u^2}{2}\Big|_{u=v}^{1-v}\right)dv = 2\int_0^{1/2} (1 - 2v)\,dv = 2(v - v^2)\Big|_0^{1/2} = \frac{1}{2}.$$

Don't forget that the Jacobian is the *absolute value* of the determinant.
A direct calculation gives us

$$\int_0^1\int_y^1 (x + y)\,dx\,dy = \int_0^1\left[\left(\frac{x^2}{2} + xy\right)\Big|_{x=y}^{1}\right]dy = \int_0^1\left(\frac{1}{2} + y - \frac{3y^2}{2}\right)dy$$

$$= \left(\frac{y}{2} + \frac{y^2}{2} - \frac{y^3}{2}\right)\Big|_0^1 = \frac{1}{2}.$$

3. (b) We compute that $x - y = 2u - 3v$ and the Jacobian is

$$\left|\frac{\partial(x, y)}{\partial(u, v)}\right| = \begin{vmatrix} 4 & 0 \\ 2 & 3 \end{vmatrix} = 12 .$$

Therefore, the integral becomes

$$12\int_0^1\int_1^2 (2u - 3v)\, dv\, du = 12\int_0^1\left[\left(2uv - \frac{3v^2}{2}\right)\Big|_{v=1}^{2}\right]du = 12\int_0^1\left(2u - \frac{9}{2}\right)du$$

$$= 12\left(u^2 - \frac{9u}{2}\right)\Big|_0^1 = 12\left(1 - \frac{9}{2}\right) = -42 .$$

Notice that we didn't even need to know what D looks like to perform the integration. By theorem 1 of section 6.2, this linear transformation maps the rectangle D* to a parallelogram D. We map the vertices and then connect them to get D.

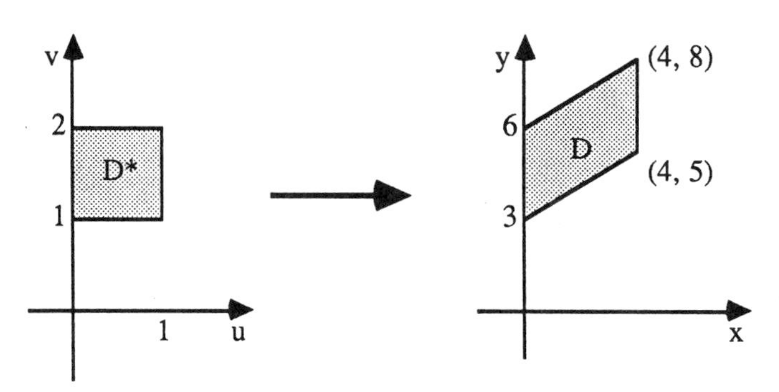

7. We compute that $x^2 + y^2 = (u^2 - v^2)^2 + 4u^2v^2 = u^4 + 2u^2v^2 + v^4 = (u^2 + v^2)^2$, and so $\sqrt{x^2 + y^2} = u^2 + v^2$. In this case, the Jacobian is

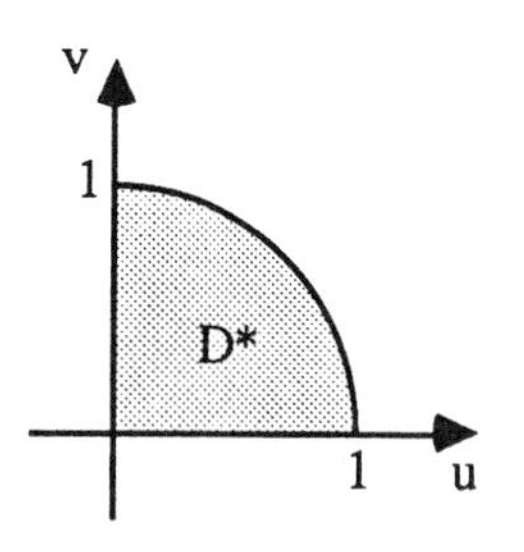

$$\left|\frac{\partial(x, y)}{\partial(u, v)}\right| = \begin{vmatrix} 2u & -2v \\ 2v & 2u \end{vmatrix} = 4u^2 + 4v^2 .$$

Therefore,

$$\int_{D^*} \frac{4(u^2 + v^2)}{u^2 + v^2}\, du\, dv = 4\int_{D^*} du\, dv .$$

This is just 4 times the area of D*, which is a quarter of a unit circle, so the answer is π.

11. The region is sketched on the next page. The transformation takes (x, y) to $(r\cos\theta, r\sin\theta)$, so the Jacobian is

$$\left|\frac{\partial(x, y)}{\partial(r, \theta)}\right| = \begin{vmatrix} \cos\theta & -r\sin\theta \\ \sin\theta & r\cos\theta \end{vmatrix} = r .$$

Recall that the area is the double integral $\int_D dx\, dy$. In polar coordinates, this becomes

$$\int_{D^*} r\, dr\, d\theta = \int_0^{2\pi}\int_0^{1 + \sin\theta} r\, dr\, d\theta = \int_0^{2\pi}\left(\frac{r^2}{2}\Big|_{r=0}^{1+\sin\theta}\right)d\theta = \frac{1}{2}\int_0^{2\pi}(1 + \sin\theta)^2\, d\theta .$$

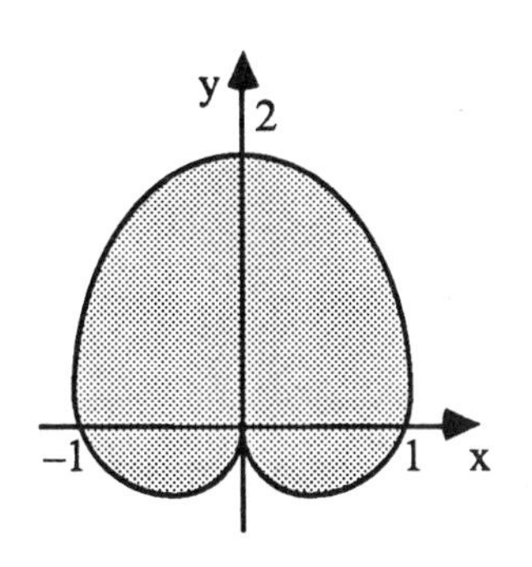

Use the half angle formula to get

$$\frac{1}{2}\int_0^{2\pi}\left(1 + 2\sin\theta + \frac{1-\cos 2\theta}{2}\right)d\theta$$

$$= \frac{1}{2}\left(\frac{3\theta}{2} - 2\cos\theta - \frac{\sin 2\theta}{4}\right)\Bigg|_0^{2\pi} = \frac{3\pi}{2}\,.$$

Compare this answer with the area formulas derived in one-variable calculus.

14. When $D^* = [0, 1] \times [0, 2\pi]$ gets mapped to a unit circle, we know we are using polar coordinates. Substitute $u = 1 + r^2$ to get

$$\int_0^{2\pi}\int_0^1 (1+r^2)^{3/2}\, r\, dr\, d\theta = \int_0^{2\pi}\int_1^2 \left(u^{3/2}\,\frac{du}{2}\right) d\theta = \int_0^{2\pi}\left(\frac{u^{5/2}}{5}\Bigg|_{u=1}^{2}\right)d\theta$$

$$= \frac{4\sqrt{2}-1}{5}\int_0^{2\pi} d\theta = \frac{2\pi(4\sqrt{2}-1)}{5}\,.$$

17. Here, we want to take the unit square and find a transformation which maps it to the given region R, a parallelogram. Note that we can map (0, 0) to (1, 0), (1, 0) to (5/2, 3/2), (0, 1) to (0, 1), and (1, 1) to (3/2, 5/2). The transformation should be linear, so we should get $(x, y) = (au + bv + c, du + ev + f)$. When $(u, v) = (0, 0)$ gets mapped to $(x, y) = (1, 0)$, we see that $c = 1$. The mapping of (1, 0) to (5/2, 3/2) gives us $a = 3/2$, and the mapping of (0, 1) to (0, 1) gives us $b = -1$. Similarly, we can determine d, e, and f. The transformation is $(3u/2 - v + 1, 3u/2 + v)$, so the Jacobian is

$$\left|\frac{\partial(x, y)}{\partial(u, v)}\right| = \begin{vmatrix} 3/2 & -1 \\ 3/2 & 1 \end{vmatrix} = 3\,.$$

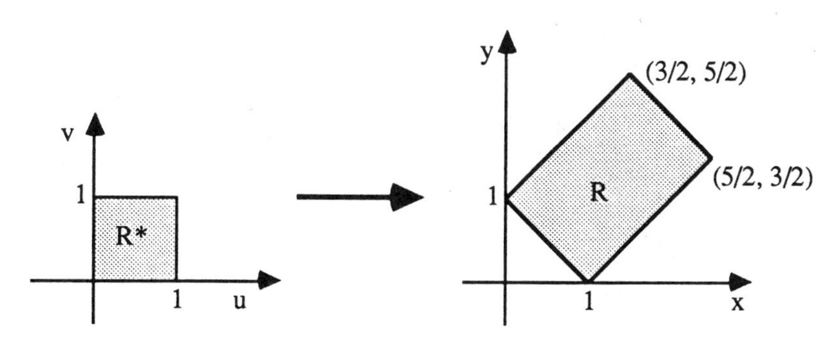

Also, we compute $x + y = 3u + 1$ and $x - y = 1 - 2v$, so the change of variables gives us

$$3\int_0^1\int_0^1 (3u+1)^2 e^{1-2v}\, du\, dv\,.$$

Since the integrand can be factored into separate factors which contain only one variable and the limits of integration are all constants, we get

$$3\left(\int_0^1 (3u+1)^2\,du\right)\left(\int_0^1 e^{1-2v}\,dv\right) = 3\left(\frac{(3u+1)^3}{9}\Big|_0^1\right)\left(\frac{e^{1-2v}}{-2}\Big|_0^1\right)$$

$$= 3\left(\frac{64-1}{9}\right)\left(\frac{e^{-1}-e^{1}}{-2}\right) = \frac{21}{2}\left(e - \frac{1}{e}\right).$$

Alternatively, let $u = x + y$, $v = x - y$. Then $1 \le u \le 4$ and $-1 \le v \le 1$, and the Jacobian is $1/2$.

23. We will use spherical coordinates. The region S can be described by

$$a \le \rho \le b,\ 0 \le \theta \le 2\pi, \text{ and } 0 \le \varphi \le \pi.$$

The Jacobian for spherical coordinates is $\rho^2 \sin\varphi$, so the integral becomes

$$\int_a^b\int_0^\pi\int_0^{2\pi} \frac{\rho^2 \sin\varphi}{\rho^3}\,d\theta\,d\varphi\,d\rho.$$

Since all of the limits of integration are constant, we can "factor" the integral as follows:

$$\left(\int_a^b \frac{d\rho}{\rho}\right)\left(\int_0^\pi \sin\varphi\,d\varphi\right)\left(\int_0^{2\pi} d\theta\right) = \ln\left(\frac{b}{a}\right)\cdot 2\cdot 2\pi = 4\pi\ln\left(\frac{b}{a}\right).$$

26. The Jacobian for cylindrical coordinates is r.

(a) After sketching B, we see that B lies over the unit disk (see below) and z extends from 0 to $\sqrt{x^2+y^2} = r$. Therefore,

$$\int_B z\,dV = \int_0^{2\pi}\int_0^1\int_0^r rz\,dz\,dr\,d\theta = \int_0^{2\pi}\int_0^1\left(\frac{rz^2}{2}\Big|_{z=0}^{r}\right)dr\,d\theta$$

$$= \int_0^{2\pi}\int_0^1 \frac{r^3}{3}\,dr\,d\theta = \int_0^{2\pi}\left(\frac{r^4}{8}\Big|_{r=0}^{1}\right)d\theta = \frac{1}{8}\int_0^{2\pi} d\theta = \frac{\pi}{4}.$$

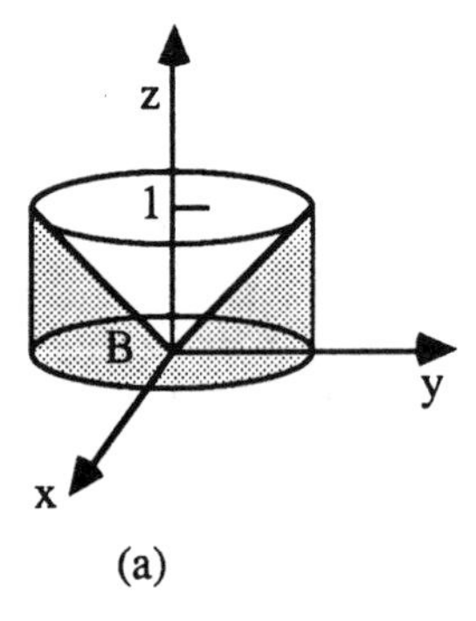

(a)

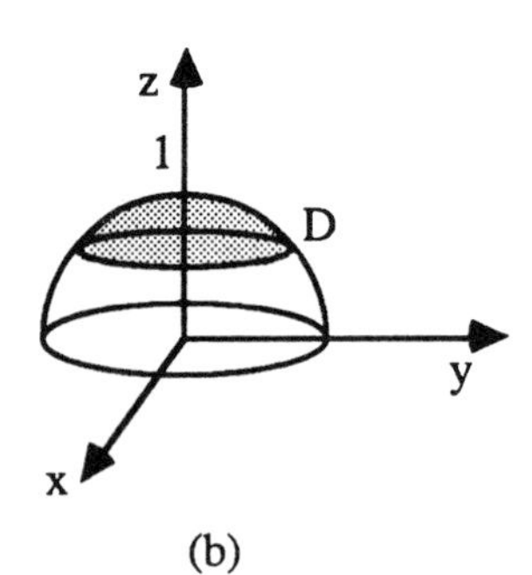

(b)

(b) The boundary $x^2 + y^2 + z^2 \le 1$ is the same as $r^2 + z^2 \le 1$ (See sketch of D above). Solving for r, we get $r \le \sqrt{1-z^2}$. Also, the integrand is $1/\sqrt{x^2+y^2+z^2} = 1/\sqrt{r^2+z^2}$. Therefore, the change of variables gives us

$$\int_0^{2\pi}\int_{1/2}^1\int_0^{\sqrt{1-z^2}} \frac{r}{\sqrt{r^2+z^2}}\, dr\, dz\, d\theta .$$

Let $u = r^2 + z^2$ to get

$$\int_0^{2\pi}\int_{1/2}^1\int_{z^2}^1 \frac{1}{2\sqrt{u}}\, du\, dz\, d\theta = \int_0^{2\pi}\int_{1/2}^1 \left(\sqrt{u}\,\Big|_{u=z^2}^{1}\right) dz\, d\theta$$

$$= \int_0^{2\pi}\int_{1/2}^1 (1-z)\, dz\, d\theta = \int_0^{2\pi}\left[\left(z - \frac{z^2}{2}\right)\Big|_{z=1/2}^{1}\right] d\theta = \frac{1}{8}\int_0^{2\pi} d\theta = \frac{\pi}{4} .$$

27. We want to find a mapping of the unit square B^* to the rectangle B. We want to map $(0, 0)$ to $(1, 0)$, $(1, 0)$ to $(4, 3)$, $(1, 1)$ to $(3, 4)$, and $(0, 1)$ to $(0, 1)$. We want a linear transformation which maps a parallelogram to a parallelogram. Our desired mapping is $(x, y) = (au + bv + c, du + ev + f)$. Mapping $(u, v) = (0, 0)$ to $(x, y) = (1, 0)$ gives us $c = 1$. Mapping $(1, 0)$ to $(4, 3)$ gives us $a = 3$. And mapping $(0, 1)$ to $(0, 1)$ gives us $b = -1$. Similarly, we can get d, e, and f. We get $(x, y) = (3u - v + 1, 3u + v)$. The Jacobian is

$$\left|\frac{\partial(x, y)}{\partial(u, v)}\right| = \begin{vmatrix} 3 & -1 \\ 3 & 1 \end{vmatrix} = 6 .$$

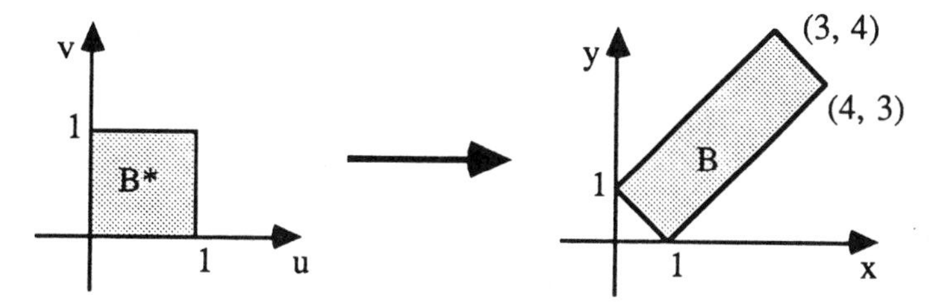

Therefore, the integral gets changed into

$$6\int_0^1\int_0^1 (6u+1)\, du\, dv = 6\int_0^1\left[(3u^2+u)\Big|_{u=0}^{1}\right] dv = 6\int_0^1 4\, dv = 24 .$$

30. In the first octant of $\mathbf{R}^3$, the spherical coordinate θ goes from 0 to $\pi/2$. The desired region of integration is sketched on the next page. Recall that the Jacobian for spherical coordinates is $\rho^2 \sin\varphi$, so the desired integral is

$$\int_0^{\pi/2}\int_{\pi/4}^{\tan^{-1}(2)}\int_0^{\sqrt{6}} \left(\frac{1}{\rho}\right)\rho^2 \sin\varphi\, d\rho\, d\varphi\, d\theta = \int_0^{\pi/2}\int_{\pi/4}^{\tan^{-1}(2)}\int_0^{\sqrt{6}} \rho \sin\varphi\, d\rho\, d\varphi\, d\theta$$

$$= \int_0^{\pi/2}\int_{\pi/4}^{\tan^{-1}(2)}\left(\frac{\rho^2}{2}\sin\varphi\,\Big|_{\rho=0}^{\sqrt{6}}\right) d\varphi\, d\theta = 3\int_0^{\pi/2}\int_{\pi/4}^{\tan^{-1}(2)} \sin\varphi\, d\varphi\, d\theta$$

$$= 3\int_0^{\pi/2}\left(-\cos\varphi\,\Big|_{\varphi=\pi/4}^{\tan^{-1}(2)}\right) d\theta .$$

From the triangle below showing $\varphi = \tan^{-1}(2)$, we get the following:

$$3\left(-\frac{1}{\sqrt{5}}+\frac{1}{\sqrt{2}}\right)\int_0^{\pi/2} d\theta = \frac{3\pi}{2}\left(\frac{1}{\sqrt{2}}-\frac{1}{\sqrt{5}}\right).$$

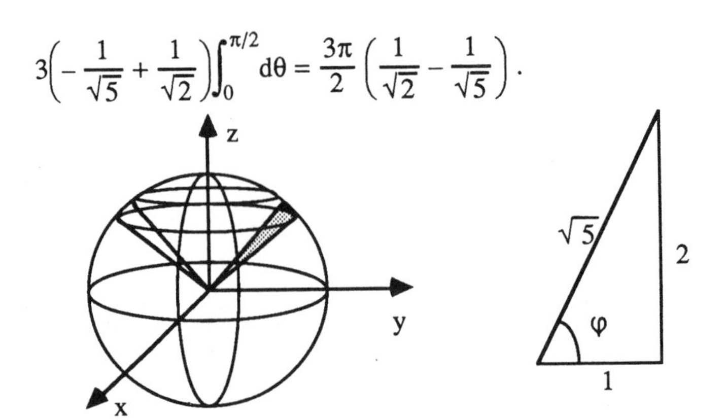

6.4 Applications of Double and Triple Integrals

GOALS

1. Be able to use double and triple integrals to compute averages, centers of mass, moments of inertia, and gravitational potential.

STUDY HINTS

1. *Averages*. In general, an average is defined to be the integral of f divided by the integral of 1. Hence, the two average formulas given in this section are

$$\frac{\int_W f\,dV}{\int_W dV} \quad \text{and} \quad \frac{\int_D f\,dA}{\int_D dA}.$$

2. *Center of mass*. This is a weighted average. In general, the center of mass of a region in $\mathbf{R}^n$ is $(\overline{x_1}, \overline{x_2}, \overline{x_3}, \ldots, \overline{x_n})$ where $\overline{x_i}$ is

$$\frac{\int_R x_i\,\rho\,dV}{\int_R \rho\,dV}$$

and $dV = dx_1\,dx_2 \cdots dx_n$ and R is the region. Since ρ is the density, $\int_R \rho\,dV$ is the mass.

3. *Moment of inertia.* A good way to remember $I_x = \int_W \rho(y^2 + z^2)\, dV$ is to notice that the integrand lacks an x term. Similarly, I_y and I_z lack a y and a z term, respectively.

4. *Gravitational potential.* Its formula is $V = GM/R$. Although you don't need to understand the physical theory behind the discussion, the mathematics should be clear to you. You probably will not need to reproduce the discussion for an exam.

5. *Geometry.* Recall that if the integrand is 1, then the triple integral $\iiint_W dV$ gives the volume of W. Similarly, the double integral $\iint_D dA$ gives the area of D.

SOLUTIONS TO SELECTED EXERCISES

3.

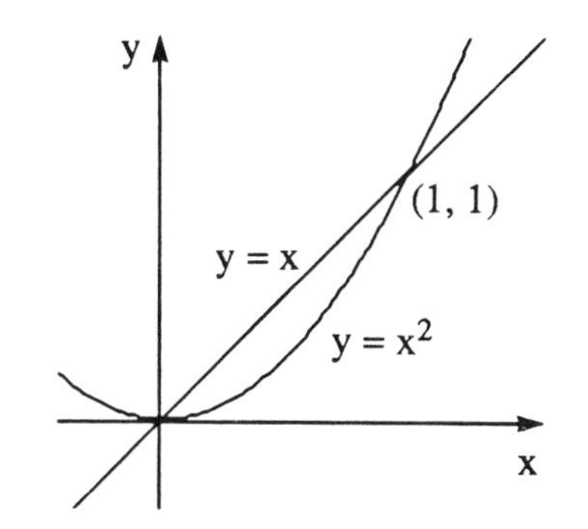

The formulas used to compute the center of mass $(\bar{x}, \bar{y})$ for a region in the plane are

$$\frac{\iint_D x\,\rho(x, y)\, dx\, dy}{\iint_D \rho(x, y)\, dx\, dy} \quad \text{and} \quad \frac{\iint_D y\,\rho(x, y)\, dx\, dy}{\iint_D \rho(x, y)\, dx\, dy}.$$

In this case, we sketch the region D and see that it can be described by

$$0 \le x \le 1 \quad \text{and} \quad x^2 \le y \le x .$$

The numerator of $\bar{x}$ is

$$\int_0^1 \int_{x^2}^{x} x(x + y)\, dy\, dx = \int_0^1 \int_{x^2}^{x} (x^2 + xy)\, dy\, dx = \int_0^1 \left[\left(x^2 y + \frac{xy^2}{2} \right) \Big|_{y=x^2}^{x} \right] dx$$

$$= \int_0^1 \left(\frac{3x^3}{2} - x^4 - \frac{x^5}{2} \right) dx = \left(\frac{3x^4}{8} - \frac{x^5}{5} - \frac{x^6}{12} \right) \Big|_0^1 = \frac{11}{120}.$$

The denominator is

$$\int_0^1 \int_{x^2}^{x} (x + y)\, dy\, dx = \int_0^1 \left[\left(xy + \frac{y^2}{2} \right) \Big|_{y=x^2}^{x} \right] dx = \int_0^1 \left(\frac{3x^2}{2} - x^3 - \frac{x^4}{2} \right) dx$$

$$= \left(\frac{x^3}{2} - \frac{x^4}{4} - \frac{x^5}{10} \right) \Big|_0^1 = \frac{18}{120}.$$

Therefore, $\bar{x}$ is $11/18$. We leave it to you to calculate that $\bar{y} = 65/126$.

6. The mass of the plate is $\int_D \rho\, dA$, which is

$$\int_0^{2\pi} \int_0^{\pi} (y^2 \sin^2 4x + 2)\, dy\, dx$$

$$= \int_0^{2\pi} \left(\frac{y^3}{3} \sin^2 4x + 2y \right) \Bigg|_0^{\pi} dx$$

$$= \int_0^{2\pi} \left(\frac{\pi^3}{3} \sin^2 4x + 2\pi \right) dx .$$

Recalling $\sin^2 t = (1 - \cos 2t)/2$, we have

$$\frac{\pi^3}{3} \int_0^{2\pi} \frac{1 - \cos 4x}{2} dx + 2\pi \int_0^{2\pi} dx$$

$$= \frac{\pi^3}{3} \left(\frac{x}{2} - \frac{\sin 4x}{8} \right) \Bigg|_0^{2\pi} + 4\pi^2 = \frac{\pi^4}{3} + 4\pi^2 .$$

The area of the plate is $2\pi \times 2\pi = 4\pi^2$. Thus the average density is mass/area , or $\pi^2/12 + 1$.

7. (a) Let ρ be the density of the box, then the mass is $\int_W \rho \, dV$, which is

$$\rho \int_0^{1/2} \int_0^1 \int_0^2 dz \, dy \, dx = \rho \left(\frac{1}{2} \right) (1)(2) = \rho .$$

(b) Again, the mass is $\int_W \rho \, dV$, which is

$$\int_0^{1/2} \int_0^1 \int_0^2 (x^2 + 3y^2 + z + 1) \, dz \, dy \, dx$$

$$= \int_0^{1/2} \int_0^1 \left[\left((x^2 + 3y^2 + 1)z + \frac{z^2}{2} \right) \Bigg|_{z=0}^{2} \right] dy \, dx$$

$$= \int_0^{1/2} \int_0^1 2(x^2 + 3y^2 + 2) \, dy \, dx = 2 \int_0^{1/2} \left[\left((x^2 + 2)y + y^3 \right) \Big|_{y=0}^{1} \right] dx$$

$$= 2 \int_0^{1/2} (x^2 + 3) \, dx = 2 \left(\frac{x^3}{3} + 3x \right) \Bigg|_0^{1/2} = 2 \left(\frac{1}{24} + \frac{3}{2} \right) = \frac{37}{12} .$$

11. The average value of f over W is $\int_W f(x, y, z) \, dV / \int_W dV$. The numerator is

$$\int_0^2 \int_0^4 \int_0^6 \sin^2 \pi z \cos^2 \pi x \, dz \, dy \, dx .$$

Since $f(x, y, z)$ can be factored so that each factor contains only one variable and the limits of integration are constants, we get

$$\left(\int_0^2 \cos^2 \pi x \, dx \right) \left(\int_0^4 dy \right) \left(\int_0^6 \sin^2 \pi z \, dz \right) .$$

Using the half-angle formulas, we get

$$\left[\int_0^2 \left(\frac{1 + \cos 2\pi x}{2} \right) dx \right] \left[\int_0^4 dy \right] \left[\int_0^6 \left(\frac{1 - \cos 2\pi z}{2} \right) dz \right] =$$

$$\frac{1}{2} \left[\left(x + \frac{\sin 2\pi x}{2\pi} \right) \Bigg|_0^2 \right] \left[y \Big|_0^4 \right] \frac{1}{2} \left[\left(z - \frac{\sin 2\pi z}{2\pi} \right) \Bigg|_0^6 \right] = \frac{1}{2} (2)(4) \frac{1}{2} (6) = 12 .$$

The denominator is just the volume of W , which is $2(4)(6) = 48$, so the average is

$12/48 = 1/4$.

14. The moment of inertia around the y axis is given by

$$I_y = \int_W \rho(x^2 + z^2)\, dx\, dy\, dz .$$

Since W is the ball of radius R, we will use the spherical coordinates (r, θ, φ) (We use r since ρ has already been used for the density.) . Since $x^2 + y^2 + z^2 = r^2$, we can rewrite the integrand as $\rho(x^2 + z^2) = \rho(r^2 - y^2)$. Thus, the moment of inertia is

$$\rho \int_0^{2\pi}\int_0^{\pi}\int_0^{R} (r^2 - r^2 \sin^2\varphi \sin^2\theta)\, r^2 \sin\varphi\, dr\, d\varphi\, d\theta$$

$$= \rho \int_0^{2\pi}\int_0^{\pi}\int_0^{R} (r^4 \sin\varphi - r^4 \sin^3\varphi \sin^2\theta)\, dr\, d\varphi\, d\theta$$

$$= \rho \int_0^{2\pi}\int_0^{\pi}\left[\left(\frac{r^5}{5}\Big|_{r=0}^{R}\right)(\sin\varphi - \sin^3\varphi \sin^2\theta)\right] d\varphi\, d\theta$$

$$= \frac{\rho R^5}{5}\int_0^{2\pi}\int_0^{\pi} (\sin\varphi - \sin^3\varphi \sin^2\theta)\, d\varphi\, d\theta .$$

To finish the integral, use the identities $\sin^2 t + \cos^2 t = 1$ and the half-angle formula: $\sin^2 t = (1 - \cos 2t)/2$:

$$= \frac{\rho R^5}{5}\int_0^{2\pi}\int_0^{\pi} [\sin\varphi - \sin\varphi(1 - \cos^2\varphi)\sin^2\theta]\, d\varphi\, d\theta$$

$$= \frac{\rho R^5}{5}\int_0^{2\pi}\left[\left(-\cos\varphi + \sin^2\theta\left(\cos\varphi - \frac{\cos^3\varphi}{3}\right)\right)\Big|_{\varphi=0}^{\pi}\right] d\theta$$

$$= \frac{\rho R^5}{5}\int_0^{2\pi}\left(2 - \frac{4}{3}\sin^2\theta\right) d\theta = \frac{\rho R^5}{5}\int_0^{2\pi}\left[2 - \frac{2}{3}(1 - \cos 2\theta)\right] d\theta$$

$$= \frac{\rho R^5}{5}\left[2\theta - \frac{2}{3}\theta + \frac{\sin 2\theta}{3}\right]\Big|_0^{2\pi} = \frac{8\rho R^5 \pi}{15} .$$

19. From the solution to example 7, $V(0, 0, R) = 4\pi G\rho(r_2^3 - r_1^3)/3R$, where r_1 and r_2 are the distances from the origin, ρ is the constant density, and $(0, 0, R)$ is a point outside the planet. Also, the solution to example 7 tells us that

$$\int_0^{\pi} \frac{\sin\varphi\, d\varphi}{\sqrt{r^2 - 2Rr\cos\varphi + R^2}} = \frac{r + R - |r - R|}{Rr} = \frac{2}{R} \text{ if } R > r .$$

When $0 \le r \le 10^4$, the gravitational potential is $4\pi G(3)[(10^4)^3 - (0)^3]/3R = (4\pi G \times 10^{12})/R$. When $10^4 \le r \le 5 \times 10^8$, the gravitational potential is

$$2\pi G \int_{10^4}^{5\times 10^8}\left[\frac{r^2\left(\dfrac{3\times 10^4}{r}\right)\left(\displaystyle\int_0^{\pi}\sin\varphi\, d\varphi\right)}{\sqrt{r^2 - 2Rr\cos\varphi + R^2}}\right] dr .$$

(another result from example 7). This is

$$2\pi G \int_{10^4}^{5\times 10^8} r^2\left(\frac{3\times 10^4}{r}\right)\left(\frac{2}{R}\right) dr = (6\times 10^4)\pi G \int_{10^4}^{5\times 10^8} \frac{2r}{R}\, dr$$

$$= (6\times 10^4)\frac{\pi G}{R} r^2 \Big|_{r=10^4}^{5\times 10^8} \approx (1.5\times 10^{22})\frac{\pi G}{R} .$$

The potential from the core of the planet is negligible, since 10^4 is a whole lot smaller than 5×10^8, so $V(0, 0, R) = (4.71\times 10^{22})G/R$.

6.5 Improper Integrals

GOALS

1. Be able to define and compute an improper integral.

STUDY HINTS

1. *Comparison.* One-variable integrals are called improper if the function becomes infinite at some point or if the limits of integration were infinite. The difference here is that a function may become infinite on a curve or a surface.

2. *Computation.* If the integral exists, just compute as usual and find a limiting value rather than substituting the limits directly. Fubini's theorem is also valid for improper integrals.

3. f *becomes infinite.* We remind you that f may become infinite between the limits of integration. In this case, we need to divide the region at the points where f is infinite. For example,

$$\int_0^1\int_{-1}^1 \frac{x}{y}\, dy\, dx = \int_0^1\int_{-1}^0 \frac{x}{y}\, dy\, dx + \int_0^1\int_0^1 \frac{x}{y}\, dy\, dx .$$

See what happens if you evaluate the antiderivative at ± 1 and not consider $y = 0$.

SOLUTIONS TO SELECTED EXERCISES

1\. (c)

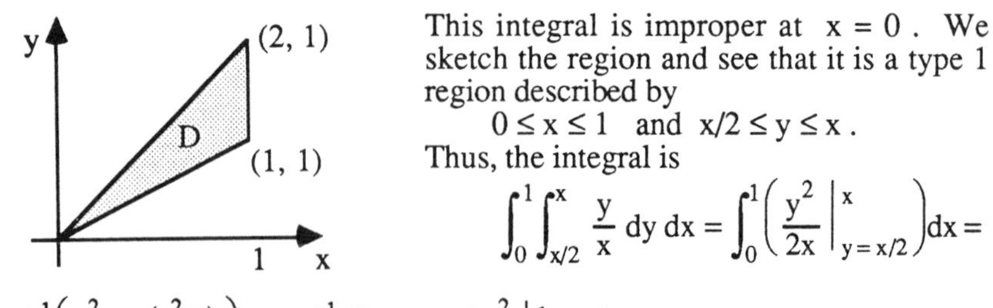

This integral is improper at $x = 0$. We sketch the region and see that it is a type 1 region described by

$$0 \le x \le 1 \text{ and } x/2 \le y \le x .$$

Thus, the integral is

$$\int_0^1\int_{x/2}^{x} \frac{y}{x}\, dy\, dx = \int_0^1\left(\frac{y^2}{2x}\Big|_{y=x/2}^{x}\right)dx =$$

$$\int_0^1\left(\frac{x^2}{2x} - \frac{(x^2/4)}{2x}\right)dx = \int_0^1 \frac{3x}{8}\, dx = \frac{3x^2}{16}\Big|_0^1 = \frac{3}{16} .$$

2. (a) Assuming D is a type 1 region, it would be reasonable to think of $b = \infty$ as an endpoint. Then we can define $\int_D f\, dA$ by

$$\int_a^\infty \int_{\varphi_1(x)}^{\varphi_2(x)} f(x, y)\, dy\, dx \ .$$

(b) The region D is like the one described in part (a). Since the limits of integration are constant, we can integrate each variable separately and then multiply the results. The integrand is factorable as follows:

$$xy \exp(-x^2 - y^2) = [x \exp(-x^2)][y \exp(-y^2)] \text{ , so}$$

$$\int_D f(x, y)\, dA = \left[\int_0^1 y \exp(-y^2)\, dy\right]\left[\int_1^\infty x \exp(-x^2)\, dx\right]$$

$$= \left(\frac{\exp(-y^2)}{-2}\Big|_0^1\right) \lim_{b\to\infty}\left(\frac{\exp(-x^2)}{-2}\Big|_1^b\right) = \left(\frac{1}{4}\right)\left(\frac{1}{e} - \frac{1}{e^2}\right).$$

5. The integrand simplifies to $1/(x+ y)$. The integral is improper wherever $y = -x$, or more specifically, the point $(0, 0)$ on D . We compute

$$\int_0^1\int_0^1 \frac{1}{x+y}\, dy\, dx = \int_0^1 \left[(\ln | x+y |) \Big|_{y=0}^1 \right] dx = \int_0^1 [\ln(1 + x) - \ln(x)]\, dx \ .$$

Integrating by parts, we get

$$[((x + 1) \ln (x + 1) - x] - (x \ln x - x)] \Big|_0^1 \ .$$

Using l'Hôpital's rule, we know that

$$\lim_{x\to 0} x \ln x = \lim_{x\to 0} \frac{\ln x}{1/x} = 0 \ ,$$

and so the integral is $2 \ln 2$.

8. When $0 \le g(x, y) \le f(x, y)$, integration over D should give us

$$\int_D 0\, dA \le \int_D g(x, y)\, dA \le \int_D f(x, y)\, dA \ .$$

However, since the integral of g is smaller than the integral of f and $\int_D g(x, y)\, dA$ does not exist, we must conclude that $\int_D f(x, y)\, dA$ also does not exist.

11. Since $x^2 + y^2 + z^2$ appears in the integrand, try using spherical coordinates. In the first octant, we have

$$0 \le \theta \le \pi/2 \text{ and } 0 \le \varphi \le \pi/2 \ .$$

Since $x^2 + y^2 + z^2 \le a^2$, we also have $0 \le \rho \le a$. Recall that the Jacobian for spherical coordinates is $\rho^2 \sin \varphi$. Substitute $\rho^2 = x^2 + y^2 + z^2$ and $\rho \cos \varphi = z$ to get

$$\int_0^{\pi/2}\int_0^{\pi/2}\int_0^a \frac{\rho^{1/2}}{\sqrt{\rho \cos \varphi + \rho^4}}\, \rho^2 \sin \varphi\, d\rho\, d\varphi\, d\theta = \int_0^{\pi/2}\int_0^{\pi/2}\int_0^a \frac{\rho^2 \sin \varphi}{\sqrt{\cos \varphi + \rho^3}}\, d\rho\, d\varphi\, d\theta \ .$$

Substitute $u = \cos \varphi + \rho^3$:

$$\int_0^{\pi/2}\int_0^{\pi/2}\int_{\cos\varphi}^{\cos\varphi + a^3} \frac{du \sin\varphi}{3\sqrt{u}}\, d\varphi\, d\theta = \int_0^{\pi/2}\int_0^{\pi/2} \frac{2}{3}\sin\varphi \left(\sqrt{u}\,\Big|_{u=\cos\varphi}^{\cos\varphi + a^3}\right) d\varphi\, d\theta$$

$$= \int_0^{\pi/2}\int_0^{\pi/2} \frac{2}{3}\sin\varphi \sqrt{\cos\varphi + a^3}\, d\varphi\, d\theta - \int_0^{\pi/2}\int_0^{\pi/2} \frac{2}{3}\sin\varphi\sqrt{\cos\varphi}\, d\varphi\, d\theta$$

$$= \int_0^{\pi/2}\left[\frac{-4}{9}(\cos\varphi + a^3)^{3/2}\Big|_{\varphi=0}^{\pi/2} + \frac{4}{9}(\cos\varphi)^{3/2}\Big|_{\varphi=0}^{\pi/2}\right] d\theta$$

$$= \frac{2\pi}{9}[(1 + a^3)^{3/2} - a^{9/2} - 1] .$$

12. Since $x^2 + y^2 + z^2$ appears in the integral, we will try to use spherical coordinates. The region D can be described by

$$1 \le r \text{ and } 0 \le \theta \le 2\pi \text{ and } 0 \le \varphi \le \pi .$$

In spherical coordinates, the integrand is $1/\rho^4$ and recall that the Jacobian is $\rho^2 \sin\varphi$. Therefore, we get

$$\int_0^{2\pi}\int_0^{\pi}\int_1^{\infty} \frac{\rho^2 \sin\varphi}{\rho^4}\, d\rho\, d\varphi\, d\theta .$$

Since all of the limits of integration are constant, we may integrate each variable separately and multiply the results:

$$\int_0^{2\pi}\int_0^{\pi}\int_1^{\infty} \frac{\sin\varphi}{\rho^2}\, d\rho\, d\varphi\, d\theta = \left(\int_0^{2\pi} d\theta\right)\left(\int_0^{\pi}\sin\varphi\, d\varphi\right)\left(\int_1^{\infty}\frac{d\rho}{\rho^2}\right)$$

$$= 2\pi\left(-\cos\varphi\,\Big|_0^{\pi}\right) \lim_{b\to\infty}\left(-\frac{1}{\rho}\Big|_1^b\right) = 2\pi(2)(1) = 4\pi .$$

6.R Review Exercises for Chapter 6

SOLUTIONS TO SELECTED EXERCISES

2. Evaluate as an iterated integral.

$$\int_0^1\int_0^x\int_0^y (y + xz)\, dz\, dy\, dx = \int_0^1\int_0^x\left[\left(yz + \frac{xz^2}{2}\right)\Big|_{z=0}^{y}\right] dy\, dx = \int_0^1\int_0^x\left(y^2 + \frac{xy^2}{2}\right) dy\, dx$$

$$= \int_0^1\left[\left(1 + \frac{x}{2}\right)\frac{y^3}{3}\Big|_{y=0}^{x}\right] dx = \int_0^1\left(\frac{x^3}{3} + \frac{x^4}{6}\right) dx = \left(\frac{x^4}{12} + \frac{x^5}{30}\right)\Big|_0^1 = \frac{7}{60} .$$

5. The plane $x + y = 1$ can be rewritten as $x = 1 - y$. Therefore, W can be described by

$$0 \le y \le 1 ,\ 0 \le x \le 1 - y ,\ \text{and } 0 \le z \le \pi .$$

Therefore, the desired integral is

$$\int_0^{\pi}\int_0^1\int_0^{1-y} x^2 \cos z\, dx\, dy\, dz = \int_0^{\pi}\int_0^1 (\cos z)\left(\frac{x^3}{3}\bigg|_{x=0}^{1-y}\right)dy\, dz$$

$$= \frac{1}{3}\int_0^{\pi}\int_0^1 (\cos z)(1-y)^3\, dy\, dz = \frac{1}{3}\int_0^{\pi}\left[(\cos z)\frac{(1-y)^4}{-4}\bigg|_{y=0}^{1}\right]dz$$

$$= \frac{1}{12}\int_0^{\pi} \cos z\, dz = \frac{1}{12}\sin z\bigg|_0^{\pi} = 0\ .$$

Alternatively, one can integrate in z first:

$$\int_0^1\int_0^{1-y}\int_0^{\pi} x^2 \cos z\, dz\, dx\, dy = \int_0^1\int_0^{1-y} (x^2 \sin z\bigg|_{z=0}^{\pi})\, dx\, dy = \int_0^1\int_0^{1-y} 0\, dx\, dy = 0\ .$$

9.

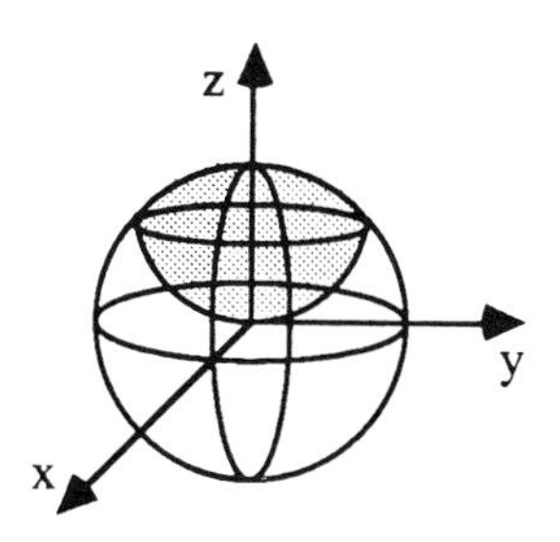

The desired volume is shown here. The surfaces intersect where $z = x^2 + y^2 = 2 - z^2$ or $z^2 + z - 2 = 0$ or $z = 1$. We disregard $z = -2$ as a solution since we want $z \geq 0$. Looking at our drawing, we can describe it in cylindrical coordinates as follows. Notice that the entire region lies over the unit disk. Also, the region lies between $z = x^2 + y^2 = r^2$ and $z = \sqrt{2 - x^2 - y^2} = \sqrt{2 - r^2}$, so the region can be described by

$$0 \leq r \leq 1\ ,\quad 0 \leq \theta \leq 2\pi\ ,\quad \text{and}\quad r^2 \leq z \leq \sqrt{2 - r^2}\ .$$

Since the Jacobian for cylindrical coordinates is r, we get

$$\int_W dv = \int_0^1\int_{r^2}^{\sqrt{2-r^2}}\int_0^{2\pi} r\, d\theta\, dz\, dr = 2\pi\int_0^1\left(rz\bigg|_{z=r^2}^{(2-r^2)^{1/2}}\right)dr$$

$$= 2\pi\int_0^1\left(r\sqrt{2-r^2} - r^3\right)dr = 2\pi\left[\frac{-(2-r^2)^{3/2}}{3} - \frac{r^4}{4}\right]\bigg|_0^1$$

$$= \frac{\pi}{3}\left(4\sqrt{2} - \frac{7}{2}\right).$$

12.

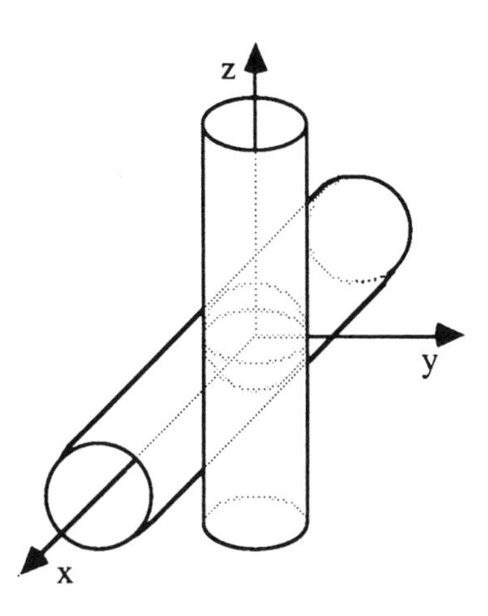

For convenience in using cylindrical coordinates, we will rotate one of the cylinders. Rather than centering one on the x axis, we will center it on the z axis. In cylindrical coordinates, the horizontal cylinder can be described by $x^2 + z^2 = 4$, or $z = \pm\sqrt{4 - r^2\cos\theta}$. The Jacobian is r. The volume is

$$\int_0^{2\pi}\int_0^2\int_{-\sqrt{4-r^2\cos^2\theta}}^{\sqrt{4-r^2\cos^2\theta}} r\, dz\, dr\, d\theta$$

$$= \int_0^{2\pi}\int_0^2\left(rz\bigg|_{z=-(4-r^2\cos^2\theta)^{1/2}}^{(4-r^2\cos^2\theta)^{1/2}}\right)dr\, d\theta$$

$$= 2\int_0^{2\pi}\int_0^2 r\sqrt{4 - r^2\cos^2\theta}\ dr\, d\theta\ .$$

Let $u = 4 - r^2\cos^2\theta$, so $r\,dr = -du/2$, and we get

$$2\int_0^{2\pi}\int_4^{4-4\cos^2\theta} \frac{\sqrt{u}}{-2}\ du\, d\theta = 2\int_0^{2\pi}\left(\frac{u^{3/2}}{-3}\Big|_{u=0}^{4-4\cos^2\theta}\right)d\theta$$

$$= \frac{-2}{3}\int_0^{2\pi} [(4 - 4\cos^2\theta)^{3/2} - \theta]d\theta = \frac{-2}{3}\int_0^{2\pi} (8\sin^3\theta - 8)\ d\theta$$

$$= \frac{-16}{3}\int_0^{2\pi} [\sin\theta(1 - \cos^2\theta) - 1]d\theta$$

$$= \frac{-16}{3}\left[-\cos\theta + \frac{\cos^3\theta}{3} - \theta\right]\Bigg|_0^{2\pi} = \frac{32\pi}{3}\ .$$

15.

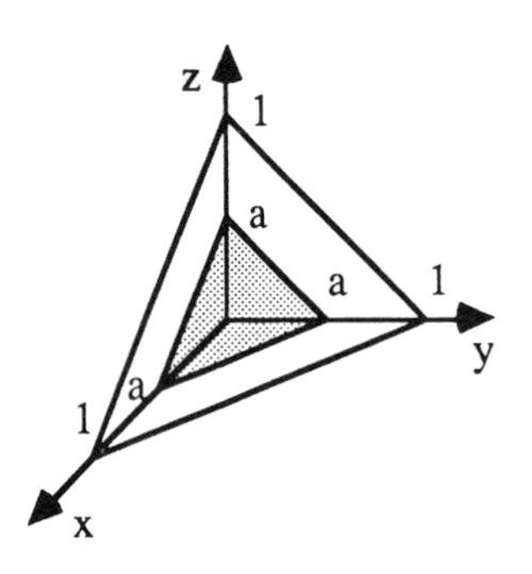

When the cut is made by the plane $x + y + z = a$, the volume of the (shaded) solid below that plane is

$$\int_0^a\int_0^{a-x}\int_0^{a-x-y} dz\, dy\, dx = \int_0^a\int_0^{a-x} (a - x - y)\, dy\, dx$$

$$= \int_0^a\left[\left((a-x)y - \frac{y^2}{2}\right)\Big|_{y=0}^{a-x}\right]dx = \int_0^a \frac{(a-x)^2}{2}\ dx\ .$$

Substitute $u = a - x$ to get

$$\frac{-1}{2}\int_{-a}^0 u^2 du = \frac{-1}{2}\left(\frac{u^3}{3}\Big|_{-a}^0\right) = \frac{a^3}{6}\ .$$

Thus, the volume for the entire solid, when $a = 1$, is $1/6$. If the solid is to be cut into n equal volumes, then the volume under $x + y + z = a$ should be $k/6n$, where k is an integer such that $1 \le k \le n - 1$. Therefore, $a^3/6 = k/6n$ implies that cuts should be made in the planes $x + y + z = (k/n)^{1/3}$.

16. (b) Dividing through by x^2 , the integrand becomes $(1/x)/[1 + (z/x)^2]$. Let $u = z/x$, so $x\,du = dz$ and the integral becomes

$$\int_0^1\int_0^y\int_0^{1/\sqrt{3}}\left(\frac{1}{1 + u^2}\right)du\, dx\, dy$$

$$= \int_0^1\int_0^y \left(\tan^{-1}u\,\Big|_{u=0}^{1/\sqrt{3}}\right) dx\, dy = \frac{\pi}{6}\int_0^1\int_0^y dx\, dy$$

$$= \frac{\pi}{6}\int_0^1 \left(x\,\Big|_{x=0}^y\right) dy = \frac{\pi}{6}\int_0^1 y\ dy = \frac{\pi}{6}\left(\frac{y^2}{2}\Big|_0^1\right) = \frac{\pi}{12}\ .$$

18. (d)

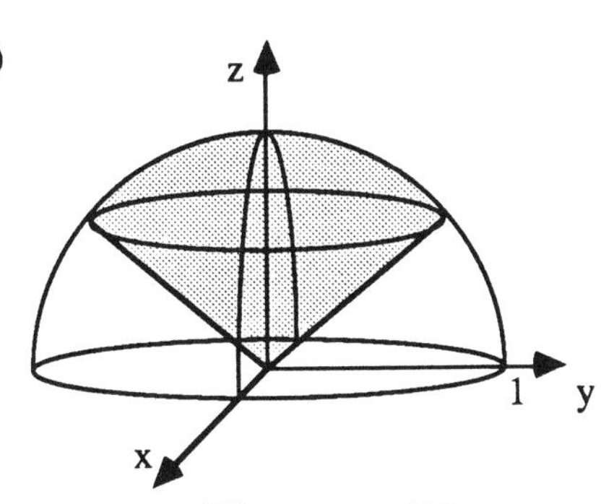

Recall that the Jacobian for spherical coordinates is $\rho^2 \sin \varphi$, and $\sin 2\varphi = 2 \sin \varphi \cos \varphi$. We factor out the Jacobian from the integrand and we are left with $2\rho \cos \varphi$, which converts in rectangular coordinates to $2z$. From our sketch of the region, we see that the region of integration is over the disk of radius 1/2, centered at the origin, so we have

$$-1/\sqrt{2} \le x \le 1/\sqrt{2} \quad \text{and} \quad -\sqrt{1/2 - x^2} \le y \le \sqrt{1/2 - x^2}.$$

Now, z extends from the cone to the sphere of radius 1. The equation of the cone is $z = x^2 + y^2$ and the upper hemisphere has the equation $z = \sqrt{1 - x^2 - y^2}$ so

$$x^2 + y^2 \le z \le \sqrt{1 - x^2 - y^2}.$$

Therefore, the integral is converted to

$$\int_{-1/\sqrt{2}}^{1/\sqrt{2}} \int_{-\sqrt{1/2 - x^2}}^{\sqrt{1/2 - x^2}} \int_{x^2 + y^2}^{\sqrt{1 - x^2 - y^2}} 2z \, dz \, dy \, dx .$$

21. (a) This improper integral is evaluated as follows:

$$\int_0^\infty \int_0^y x \exp(-y^3) \, dx \, dy = \int_0^\infty \left(\frac{x^2}{2} \Big|_{x=0}^{y} \right) \exp(-y^3) \, dy$$

$$= \frac{1}{2} \int_0^\infty y^2 \exp(-y^3) \, dy = \lim_{b \to \infty} \left(\frac{\exp(-y^3)}{-6} \Big|_0^b \right) = \frac{1}{6} .$$

24. (a) Since we are integrating a spherical region, we will use spherical coordinates and the Jacobian $\rho^2 \sin \varphi$.

$$\iiint_D (x^2 + y^2 + z^2) xyz \, dV$$

$$= \int_0^{2\pi} \int_0^{\pi} \int_0^R \rho^2 (\rho \cos\theta \sin \varphi)(\rho \sin \theta \sin \varphi)(\rho \cos \varphi) \rho^2 \sin \varphi \, d\rho \, d\varphi \, d\theta$$

$$= \int_0^{2\pi} \int_0^{\pi} \int_0^R \rho^7 \cos \theta \sin \theta \cos \varphi \sin^3 \varphi \, d\rho \, d\varphi \, d\theta .$$

Since all of the limits of integration are constants, we can integrate each variable separately and multiply the results. Thus, the integral is

$$\left(\int_0^R \rho^7 \, d\rho \right) \left(\int_0^{2\pi} \cos \theta \sin \theta \, d\theta \right) \left(\int_0^{\pi} \cos \varphi \sin^3 \varphi \, d\varphi \right)$$

$$= \left(\frac{\rho^8}{8} \Big|_0^R \right) \left(\frac{\sin^2 \theta}{2} \Big|_0^{2\pi} \right) \left(\frac{\sin^4 \varphi}{4} \Big|_0^{\pi} \right) = 0 .$$

27. Let $u = x + y$, $v = y - x$, or $x = (u - v)/2$, $y = (u + v)/2$. The reader should verify that the Jacobian $|\partial(u, v)/\partial(x, y)|$ is 1/2. From the figures, $0 \le u \le 1$ and $-u \le v \le u$.

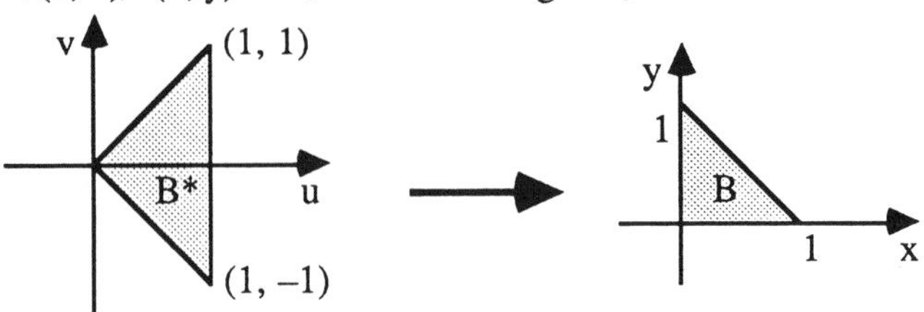

Then,

$$\int\int_B \exp\left[\frac{y-x}{y+x}\right] dx\,dy$$

$$= \int_0^1 \int_{-u}^{u} \frac{\exp(v/u)}{2}\, dv\, du = \frac{1}{2}\int_0^1 u\,(e^{v/u}\Big|_{-u}^{u})\, du$$

$$= \frac{1}{2}\left(e - \frac{1}{e}\right)\int_0^1 u\, du = \frac{1}{4}\left(e - \frac{1}{e}\right).$$

30. The total mass is the integral of the density. Since we're integrating over a sphere, use spherical coordinates. The sphere of radius R is described by

$$0 \le \rho \le R\,,\ 0 \le \varphi \le \pi\,, \text{ and } 0 \le \theta \le 2\pi\,.$$

The distance d from the origin is the same as ρ and the Jacobian is $\rho^2 \sin\varphi$. Therefore, the total mass is

$$\int_0^{2\pi}\int_0^{\pi}\int_0^{R} \frac{\rho^2 \sin\varphi}{1+\rho^3}\, d\rho\, d\varphi\, d\theta\,.$$

Since all of the limits of integration are constants, we can integrate each variable separately and multiply the results. We get

$$\left(\int_0^{2\pi} d\theta\right)\left(\int_0^{\pi} \sin\varphi\, d\varphi\right)\left(\int_0^{R} \frac{\rho^2}{1+\rho^3}\, d\rho\right)$$

$$= (2\pi)[-\cos\varphi\Big|_0^{\pi}]\left[\frac{1}{3}\ln(1+\rho^3)\Big|_0^{R}\right] = \frac{4\pi}{3}\ln(1+R^3)\,.$$

33. (a) Let $T(x, y, z)$ be the temperature at (x, y, z). Then the average temperature is $\int_C T\,dV / \int_C dV$. By definition, $d = \sqrt{x^2 + y^2 + z^2}$, so $d^2 = x^2 + y^2 + z^2$ and $T(x, y, z) = 32(x^2 + y^2 + z^2)$. The numerator is

$$32\int_{-1}^{1}\int_{-1}^{1}\int_{-1}^{1} (x^2 + y^2 + z^2)\, dx\, dy\, dz$$

$$= 32\int_{-1}^{1}\int_{-1}^{1}\left[\left(\frac{x^3}{3} + (y^2 + z^2)x\right)\Big|_{x=-1}^{1}\right] dy\, dz$$

$$= 64\int_{-1}^{1}\int_{-1}^{1}\left(\frac{1}{3} + y^2 + z^2\right) dy\, dz$$

$$= 64\int_{-1}^{1}\left[\left(\left(\frac{1}{3}+z^2\right)y+\frac{y^3}{3}\right)\Big|_{y=-1}^{1}\right]dz$$

$$= 128\int_{-1}^{1}\left(\frac{2}{3}+z^2\right)dz = 128\left(\frac{2z}{3}+\frac{z^3}{3}\right)\Big|_{-1}^{1} = 256\,.$$

The denominator is just the volume of the cube C, which is 8, so the average temperature on C is $256/8 = 32$.

(b) The average temperature is attained wherever $32d^2 = 32$ or $d = 1$. Thus, the average temperature is attained at all points on a sphere of radius 1, centered at the origin.

34. The inequality $y^2 + z^2 < 1/4$ describes the inside of a circular cylinder of radius 1/2, centered on the x axis. The inequality $(x-1)^2 + y^2 + z^2 \le 1$ is a ball of radius 1 centered at $(1, 0, 0)$. We also want $x \ge 1$, so we get the region shown. By symmetry, we easily see that $\bar{y} = \bar{z} = 0$.
For convenience, we will shift the region in question so that its axis of symmetry is the z^* axis and we will place the ball's center at the origin. For the center of mass of the new region D^*, it is the integral of z^* over D^* divided by the volume of D^*. D^* is described by the cylindrical coordinates

$$0 \le r \le 1/2\,,\ 0 \le \theta \le 2\pi\,,\ \text{and}\ \ 0 \le z^* \le \sqrt{1-r^2}\ .$$

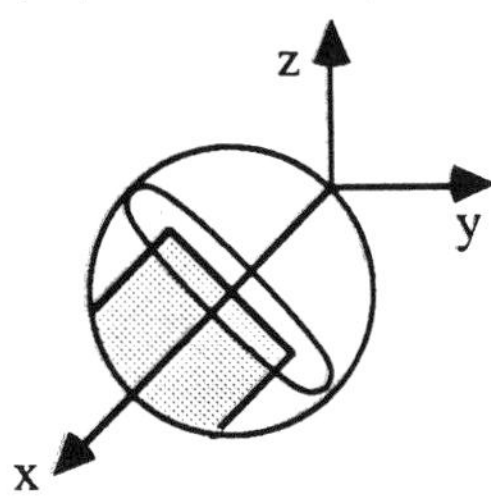

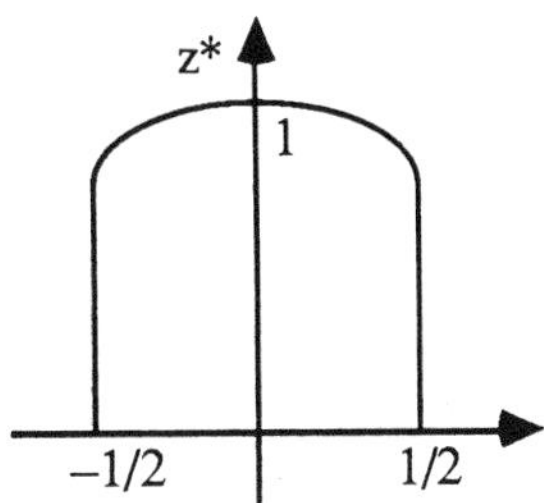

The Jacobian is r, so the integral of z^* over D^* is

$$\int_0^{2\pi}\int_0^{1/2}\int_0^{\sqrt{1-r^2}} rz^*\,dz^*\,dr\,d\theta = \int_0^{2\pi}\int_0^{1/2}\left(\frac{r(z^*)^2}{2}\Big|_{z^*=0}^{\sqrt{1-r^2}}\right)dr\,d\theta$$

$$= \int_0^{2\pi}\int_0^{1/2}\left(\frac{r-r^3}{2}\right)dr\,d\theta$$

$$= \frac{1}{2}\int_0^{2\pi}\left[\left(\frac{r^2}{2}-\frac{r^4}{4}\right)\Big|_{r=0}^{1/2}\right]d\theta = \frac{7}{128}\int_0^{2\pi}d\theta = \frac{7\pi}{64}\ .$$

The volume of D^* is

$$\int_0^{2\pi}\int_0^{1/2}\int_0^{\sqrt{1-r^2}} r\,dz^*\,dr\,d\theta = \int_0^{2\pi}\int_0^{1/2}\left(rz^*\Big|_{z^*=0}^{\sqrt{1-r^2}}\right)dr\,d\theta$$

$$= \int_0^{2\pi}\int_0^{1/2} r\sqrt{1-r^2}\,dr\,d\theta = \int_0^{2\pi}\left[\frac{-(1-r^2)^{3/2}}{3}\Big|_{r=0}^{1/2}\right]d\theta$$

$$= \left(\frac{1}{3} - \frac{\sqrt{3}}{8}\right)\int_0^{2\pi} d\theta = 2\pi\left(\frac{1}{3} - \frac{\sqrt{3}}{8}\right).$$

D* was shifted down 1/2 unit, so for the original region, $\overline{x}$ is

$$\frac{1}{2} + \frac{7\pi/64}{2\pi\left(\frac{1}{3} - \frac{\sqrt{3}}{8}\right)} = \frac{1}{2} + \frac{7}{128\left(\frac{1}{3} - \frac{\sqrt{3}}{8}\right)}.$$

39. Notice that $x^2 + y^2$ occurs in the integrand, so we will try to use polar coordinates with Jacobian r. Let $\mathbf{R}^2$ be a disk with infinite radius, so $\mathbf{R}^2$ is described by

$$0 \le r \text{ and } 0 \le \theta \le 2\pi.$$

Thus,

$$\int_{\mathbf{R}^2} f(x, y)\, dA = \int_0^{2\pi}\int_0^{\infty} \frac{r}{(1 + r^2)^{3/2}}\, dr\, d\theta$$

$$= \int_0^{2\pi}\left[-\lim_{b \to \infty} (1 + r^2)^{-1/2}\Big|_{r=0}^{b}\right] d\theta = \int_0^{2\pi} d\theta = 2\pi.$$

CHAPTER 7
INTEGRALS OVER PATHS AND SURFACES

7.1 The Path Integral

GOALS

1. Be able to compute a path integral.

STUDY HINTS

1. *Notation.* The path integral is denoted by $\int_{\boldsymbol{\sigma}} f\, ds$, where $ds = \|\boldsymbol{\sigma}'(t)\|\, dt$ and $\boldsymbol{\sigma}(t)$ is a path.

2. *Definition.* The path integral is an integration of *real-valued* functions performed over a curve. The integrand f only needs to be defined on $\boldsymbol{\sigma}$ and f must be piecewise continuous on $\boldsymbol{\sigma}$. In addition, $\boldsymbol{\sigma}$ needs to be piecewise continuous differentiable. The integration of *vector-valued* functions will be discussed in the next section.

3. *Computation.* If $\boldsymbol{\sigma}(t)$ is defined for $a \le t \le b$ and f is defined on $\boldsymbol{\sigma}$, then the path integral is

$$\int_a^b f(\boldsymbol{\sigma}(t))\, \|\boldsymbol{\sigma}'(t)\|\, dt\,.$$

4. *Relation to arc length.* If $f = 1$, then the path integral is the arc length formula of section 3.2.

SOLUTIONS TO SELECTED EXERCISES

2. (b) The path integral is defined by

$$\int_a^b f(\boldsymbol{\sigma}(t))\, \|\boldsymbol{\sigma}'(t)\|\, dt\,.$$

In this case, $f(\boldsymbol{\sigma}(t)) = \cos t$, $\boldsymbol{\sigma}'(t) = (\cos t, -\sin t, 1)$, and $\|\boldsymbol{\sigma}'(t)\| = [\cos^2 t + \sin^2 t + 1]^{1/2} = \sqrt{2}$. Therefore the path integral is

$$\int_0^{2\pi} \cos t \cdot \sqrt{2}\, dt = \sqrt{2} \sin t \Big|_0^{2\pi} = 0\,.$$

3. (b) Let $x = t$, $y = 3t$, and $z = 2t$. Here, $f = yz = (3t)(2t)$ and $ds = \|\boldsymbol{\sigma}'(t)\| = [1^2 + 3^2 + 2^2]^{1/2} = \sqrt{14}$. Then the path integral is

$$\int_{\sigma} f\, ds = \int_1^3 (3t)(2t)\sqrt{14}\, dt = 6\sqrt{14}\int_1^3 t^2\, dt = 2\sqrt{14}\, t^3 \Big|_1^3 = 52\sqrt{14}\,.$$

4. (a) Recall that in polar coordinates,

$$x = r(\theta)\cos\theta \quad \text{and} \quad y = r(\theta)\sin\theta\,.$$

Here we treat r as a function of θ, and so

$$dx = (r'(\theta)\cos\theta - r(\theta)\sin\theta)\, d\theta \text{ and}$$
$$dy = (r'(\theta)\sin\theta + r(\theta)\cos\theta)\, d\theta\,.$$

Also, we know that

$$\begin{aligned} ds &= \sqrt{dx^2 + dy^2}\, d\theta \\ &= \{[r'(\theta)\cos\theta - r(\theta)\sin\theta]^2 + [r'(\theta)\sin\theta + r(\theta)\cos\theta]^2\}^{1/2}\, d\theta \\ &= [(r'(\theta))^2(\cos^2\theta + \sin^2\theta) + (r(\theta))^2(\cos^2\theta + \sin^2\theta)]^{1/2}\, d\theta \\ &= \sqrt{r^2 + (dr/d\theta)^2}\, d\theta\,. \end{aligned}$$

Hence the path integral of $f(x, y) = f(r\cos\theta, r\sin\theta)$ is

$$\int_{\sigma} f(x, y)\, ds = \int_{\theta_1}^{\theta_2} f(r\cos\theta, r\sin\theta)\sqrt{r^2 + \left(\frac{dr}{d\theta}\right)^2}\, d\theta\,.$$

7. (a) From $t = -1$ to $t = 0$, the path is a straight line going from (1, 1, 1) to (0, 0, 0); from $t = 0$ to $t = 1$, the path is a straight line going from (0, 0, 0) to (1, 1, 1), so actually the same path was covered twice. Hence, the path integral is

$$\int_a^b f(x(t), y(t))\sqrt{(x'(t))^2 + (y'(t))^2}\, dt = 2\int_0^1 (2t^4 - t^4)\sqrt{(4t^3)^2 + (4t^3)^2}\, dt$$
$$= 2\int_0^1 4t^4\sqrt{2t^6}\, dt = 8\sqrt{2}\int_0^1 t^7\, dt = \sqrt{2}\,.$$

The answer could be thought of as the area that needs to be covered if one wants to paint the two sides of a fence erected along the path $x = t^4$, $y = t^4$, of height $f(x(t), y(t))$.

(b) The arc length function is the path integral when $f = 1$, so it is

$$\int_a^b f(\boldsymbol{\sigma}(\tau))\, \|\boldsymbol{\sigma}'(\tau)\|\, d\tau\,.$$

where a is the starting point. To evaluate $s(t)$, we need to break up the interval into $[-1, 0]$ and $[0, 1]$. For $-1 \le t \le 0$, $\|\boldsymbol{\sigma}'(\tau)\| = [(4\tau^3)^2 + (4\tau^3)^2]^{1/2} = 4\sqrt{2}\,|\tau|^3 = -4\sqrt{2}\,\tau^3$ since we also have $-1 \le \tau \le 0$. Then

$$s(t) = -\int_{-1}^t 4\sqrt{2}\,\tau^3\, d\tau = -\sqrt{2}(t^4 - 1)\,.$$

So $t^4 = -s/\sqrt{2} + 1$ and $4t^3\, dt = -ds/\sqrt{2}$. When $t = -1$, $s = 0$; and when $t = 0$, $s = \sqrt{2}$. For $0 \le t \le 1$, $\|\boldsymbol{\sigma}'(\tau)\| = [(4\tau^3)^2 + (4\tau^3)^2]^{1/2} = 4\sqrt{2}\,\tau^3$.

We also need to add the length from -1 to 0, so we get, for $0 \le t \le 1$,

$$s(t) = \sqrt{2} + \int_0^t 4\sqrt{2}\,\tau^3\,d\tau = \sqrt{2}(t^4 + 1)\,.$$

So $t^4 = s/\sqrt{2} - 1$ and $4t^3\,dt = ds/\sqrt{2}$. When $t = 0$, $s = \sqrt{2}$; and when $t = 1$, $s = 0$. Putting all this together, the path integral becomes

$$\int_0^{\sqrt{2}} \left(-\frac{s}{\sqrt{2}} + 1\right) ds + \int_{\sqrt{2}}^0 \left(\frac{s}{\sqrt{2}} - 1\right) ds$$

$$= \left(-\frac{s^2}{2\sqrt{2}} + s\right)\Bigg|_0^{\sqrt{2}} + \left(\frac{s^2}{2\sqrt{2}} - s\right)\Bigg|_{\sqrt{2}}^0 = \sqrt{2}\,.$$

10. (a) When density is a constant k, we have

Mass $= k \times$ length of wire .

The length of the wire is

$$\int_a^b \|\mathbf{p}'(\theta)\|\; d\theta\,,$$

so the mass is

$$2\int_0^\pi \sqrt{0^2 + (a\cos\theta)^2 + (-a\sin\theta)^2}\; d\theta = 2\int_0^\pi a\,d\theta = 2a\pi\,.$$

(b) Recall that a coordinate of the center of mass, $\overline{x_i}$, is

$$\frac{\sum m_i x_i}{\sum m_i}\,.$$

Here, we have a thin wire in the yz plane. By symmetry, $\overline{x} = \overline{z} = 0$. In this case, $y = a\sin\theta$, so the center of mass for y, with $k = 2$, is

$$\frac{\int k\,y\,ds}{\int k\,ds} = \frac{\int_0^\pi a\sin\theta \cdot 2a\,d\theta}{2a\pi} = \frac{a}{\pi}\int_0^\pi \sin\theta\,d\theta = \frac{2a}{\pi}\,.$$

Thus, the center of mass is $(0, 2a/\pi, 0)$. Note that the center of mass is *not* on the wire.

13. The trick of this exercise is to use the correct coordinate systems. The intersection of the sphere and the plane is a unit circle, but in an unfamiliar orientation. The unit normal vector of the plane $x + y + z = 0$ is $(\mathbf{i} + \mathbf{j} + \mathbf{k})/\sqrt{3}$, so we need to find two orthogonal vectors in the plane $x + y + z = 0$. For example, $(1, 0, -1)$ and $(1, -2, 1)$ are a pair. Normalization yields $\mathbf{u} = (1, 0, -1)/\sqrt{2}$ and $\mathbf{v} = (1, -2, 1)/\sqrt{6}$. The circle is now parametrized as $\boldsymbol{\sigma}(\theta) = (x, y, z) = \cos\theta\,\mathbf{u} + \sin\theta\,\mathbf{v}$, $0 \le \theta < 2\pi$. Take the first component of $\boldsymbol{\sigma}$ to get $x = (\cos\theta)/\sqrt{2} + (\sin\theta)/\sqrt{6}$. Then the mass is

$$\int_C x^2\, ds = \int_0^{2\pi} \left(\frac{1}{\sqrt{2}} \cos\theta + \frac{1}{\sqrt{6}} \sin\theta \right)^2 \|\boldsymbol{\sigma}'(\theta)\|\, d\theta$$

$$= \int_0^{2\pi} \left(\frac{1}{\sqrt{2}} \cos\theta + \frac{1}{\sqrt{6}} \sin\theta \right)^2 \cdot 1\, d\theta$$

$$= \int_0^{2\pi} \left(\frac{1}{2}\cos^2\theta + \frac{2\sin\theta\cos\theta}{\sqrt{12}} + \frac{1}{6}\sin^2\theta \right) d\theta\ .$$

You should verify that $\|\boldsymbol{\sigma}'(t)\| = 1$. Now use the half-angle formulas to get

$$\left[\frac{1}{2}\left(\frac{\theta}{2} + \frac{\sin\theta}{2}\right) + \frac{\sin^2\theta}{\sqrt{12}} + \frac{1}{6}\left(\frac{\theta}{2} - \frac{\sin\theta}{2}\right) \right] \Bigg|_0^{2\pi} = 2\pi/3.$$

7.2 Line Integrals

GOALS

1. Be able to compute a line integral.
2. Be able to explain the difference between a line integral and a path integral.

STUDY HINTS

1. *Definition.* The line integral deals with vector fields, whereas the path integral dealt with scalar functions. It is required that $\boldsymbol{\sigma}$ be C^1 and $\mathbf{F}$ be continuous on $\boldsymbol{\sigma}$ (or at least piecewise continuous for both $\boldsymbol{\sigma}$ and $\mathbf{F}$).
2. *Physical interpretation.* The line integral is most commonly associated with the work performed by $\mathbf{F}$ along a path.
3. *Computation.* If $\mathbf{F}$ is defined on $\boldsymbol{\sigma}$ and $\boldsymbol{\sigma}$ is defined for $a \le t \le b$, then the line integral is $\int_a^b \mathbf{F}(\boldsymbol{\sigma}(t)) \cdot \boldsymbol{\sigma}'(t)\, dt$. Note that the integrand is a dot product, so the integral is a scalar. The formula involving $\mathbf{T}(t)$, the unit tangent, may be more difficult to use since you need to compute $\mathbf{T}(t)$ as well as $\|\boldsymbol{\sigma}'(t)\|$.
4. *Sign interpretation.* Positive work means that the force field did a net amount of work on the object; such is the case when the motion of the object is in the direction of the force. If work is negative, then this amount of work is done by the object on the force field.

5. *Reparametrization and orientation.* If σ is reparametrized and the orientation is preserved, the value of a line integral does not change. If the orientation is reversed, only the sign of the value of the line integral changes. You can substitute for the endpoints to be sure the direction is correct. Orientation is not important with the path integral of the previous section. Reparametrization allows us to break up a curve and integrate over each segment separately and with convenient parametrizations (see example 11). Be sure the path is traversed the correct number of times.

6. *Gradients.* The line integral of a gradient field depends only on the endpoints. We will see how to make use of this fact in section 8.3. Given this fact, we can see that the line integral of a gradient along a closed path is 0.

SOLUTIONS TO SELECTED EXERCISES

1. (a) The line integral of $\mathbf{F}$ along σ is

$$\int_0^1 \mathbf{F}(\sigma(t)) \cdot \sigma'(t)\, dt .$$

We have $\sigma(t) = (t, t, t)$, so $\sigma'(t) = (1, 1, 1)$ and $\mathbf{F}(\sigma(t)) = (t, t, t)$. Therefore, the desired line integral is

$$\int_0^1 (t\mathbf{i} + t\mathbf{j} + t\mathbf{k}) \cdot (\mathbf{i} + \mathbf{j} + \mathbf{k})\, dt = \int_0^1 (t + t + t)\, dt = \frac{3}{2} t^2 \Big|_0^1 = \frac{3}{2} .$$

2. (c) We need to break up the integral into two parts. For the line segment joining $(1, 0, 0)$ to $(0, 1, 0)$, the easiest way to parametrize this line is to find $\mathbf{l}(t) = \mathbf{u}(t) + t\mathbf{v}(t)$ such that $\mathbf{l}(0) = (1, 0, 0)$ and $\mathbf{l}(1) = (0, 1, 0)$. Letting $\mathbf{u}(t) = (1, 0, 0)$ and $\mathbf{v}(t) = (0, 1, 0) - (1, 0, 0) = (-1, 1, 0)$ gives us $\mathbf{l}_1(t) = (1, 0, 0) + t(-1, 1, 0)$ or $(x, y, z) = (1 - t, t, 0)$ for $0 \le t \le 1$, so $dx = -dt$, $dy = dt$, and $dz = 0$. Substitute these values to get

$$\int_{\mathbf{l}_1} (yz\, dx + xz\, dy + xy\, dz) = \int_0^1 [t \cdot 0 \cdot (-dt) + (1 - t) \cdot 0 \cdot dt + (1 - t) \cdot t \cdot 0] = 0 .$$

For the line segment joining $(0, 1, 0)$ to $(0, 0, 1)$, use the same method. We find $\mathbf{l}_2(t) = (0, 1, 0) + t(0, -1, 1)$ or $(x, y, z) = (0, 1 - t, t)$ for $0 \le t \le 1$, so $dx = 0$, $dy = -dt$, and $dz = dt$. Substitute these values to get

$$\int_{\mathbf{l}_2} (yz\, dx + xz\, dy + xy\, dz) = \int_0^1 [(1 - t) \cdot t \cdot 0 + 0 \cdot t \cdot (-dt) + 0 \cdot (-t)\, dt] = 0 .$$

Adding the two line integrals together, we get $0 + 0 = 0$.

4. (a) By definition, we have

$$\int_\sigma \mathbf{F} \cdot d\mathbf{s} = \int_\sigma \mathbf{F}(\sigma(t)) \cdot \sigma'(t)\, dt .$$

We are given that $\mathbf{F}$ and σ' are perpendicular, so $\mathbf{F} \cdot \sigma'(t) = 0$. Therefore, the line integral is 0.

(b) Again, we use the definition of the line integral. In addition, we use the definition of the dot product to get

$$\int_\sigma \mathbf{F} \cdot d\mathbf{s} = \int_\sigma \mathbf{F}(\sigma(t)) \cdot \sigma'(t)\, dt = \int_\sigma \|\mathbf{F}(\sigma(t))\|\, \|\sigma'(t)\| \cdot \cos(0)\, dt = \int_\sigma \|\mathbf{F}\|\, ds\,.$$

This results because the cosine between parallel vectors is 0 and by definition, $ds = \|\sigma'(t)\|\, dt$.

7. We are given that $\sigma(t) = (x, y, z) = (t, t^n, 0)$, $0 \le t \le 1$, where $n = 1, 2, 3, \ldots$. Differentiation of each component gives us $\sigma'(t) = (1, nt^{n-1}, 0)$, so $dx = dt$, $dy = nt^{n-1}$, and $dz = 0$. Therefore, $\int_\sigma [y\, dx + (3y^3 - x)dy + z\, dz] =$

$$\int_0^1 (t^n\, dt + (3t^{3n} - t)nt^{n-1}\, dt + 0) = \int_0^1 (t^n + 3nt^{4n-1} - nt^n)\, dt =$$

$$\left(\frac{1-n}{1+n} t^{n+1} + \frac{3}{4} n^{4n}\right)\Big|_0^1 = \frac{1-n}{1+n} + \frac{3}{4}\,.$$

11. Given that $\sigma(t)$ is a path and $\mathbf{T}$ is a unit tangent, we have

$$\mathbf{T} = \frac{\sigma'(t)}{\|\sigma'(t)\|}$$

by definition. Since $\mathbf{v} \cdot \mathbf{v} = \|\mathbf{v}\|^2$, we get the line integral

$$\int_\sigma \mathbf{T} \cdot d\mathbf{s} = \int_\sigma \frac{\sigma'(t)}{\|\sigma'(t)\|} \cdot \sigma'(t)\, dt = \int_\sigma \frac{\|\sigma'(t)\|^2}{\|\sigma'(t)\|}\, dt = \int_\sigma \|\sigma'(t)\|\, dt\,.$$

The last integral is simply the arc length of $\sigma(t)$.

14. If a vector field is a gradient, then the integral should only depend on the endpoints. For a closed curve, the starting point is the same as the ending point, so theorem 3 tells us that the integral is 0.

7.3 Parametrized Surfaces

GOALS

1. Be able to parametrize a given surface.
2. Be able to determine if a surface is smooth and/or differentiable.
3. Be able to compute a tangent plane for parametrized surfaces.

STUDY HINTS

1. *Parametrized surfaces.* Recall that curves in the plane could be parametrized by two functions of one variable, namely, $x = f(t)$ and $y = g(t)$. Now, we extend this idea to surfaces. We do this by letting x, y, and z be functions of two variables, i.e, now we have three functions of two variables, $x = f(u, v)$, $y = g(u, v)$, and $z = h(u, v)$.

2. *Differentiable surfaces.* Such a surface is parametrized by $(f(u, v), g(u, v), h(u, v))$ and f, g, and h are all differentiable functions.

3. *Review.* Recall that a vector $\mathbf{n}$ which is normal to a plane gives the coefficients of the plane equation. See section 1.3.

4. *Tangent vectors.* Like a partial derivative, we can hold u constant to get $\mathbf{T}_v = (f_v, g_v, h_v)$. Likewise, holding v constant gives us $\mathbf{T}_u = (f_u, g_u, h_u)$.

5. *Tangent plane.* By holding either parameter, u or v, constant, we get a curve on the surface and if the curve is "nice" , it will have a tangent line. Similarly, we can get another tangent line by holding the other parameter constant. Those two tangent lines determine the tangent plane, and the normal vector to the plane is $\mathbf{T}_u \times \mathbf{T}_v$.

6. *Smooth.* A surface is said to be smooth if $\mathbf{T}_u \times \mathbf{T}_v \neq \mathbf{0}$. When the cross product is not $\mathbf{0}$, a tangent plane exists and consequently, the surface has no pointed regions (like a cone). Differentiability does not imply smooth (see example 1), but smoothness requires differentiability for otherwise $\mathbf{T}_u$ and $\mathbf{T}_v$ would not be defined.

7. *Important parametrizations.* Know the following parametrizations:
   (a) Circle of radius r : $x = r\cos\theta$, $y = r\sin\theta$, $0 \leq \theta \leq 2\pi$.
   (b) The hyperbola $x^2 - y^2 = 1$: $x = \cosh t$, $y = \sinh t$.

(c) Sphere of radius ρ : $x = \rho \cos\theta \sin\varphi$, $y = \rho \sin\theta \sin\varphi$, $z = \rho \cos\varphi$, $0 \le \theta < 2\pi, 0 \le \varphi < \pi$. This is the same as the definition of spherical coordinates.

(d) A surface $z = g(x, y)$: $x = u, y = v, z = g(u, v)$.

8. *Normal to a graph.* If $z = f(x, y)$, the normal to the graph is $\left(-\frac{\partial f}{\partial x}, -\frac{\partial f}{\partial y}, 1\right)$. This is useful to know for the coming sections. Note that if we let $g = z - f(x, y)$, then the (unnormalized) normal to the surface is $\mathbf{n} = \nabla g$.

SOLUTIONS TO SELECTED EXERCISES

3. The normal to the desired tangent plane is $\mathbf{T}_u \times \mathbf{T}_v$. We compute $\mathbf{T}_u = (\partial x/\partial u, \partial y/\partial u, \partial z/\partial u) = (2u, \sin e^v, (1/3)\cos e^v)$. Similarly, $\mathbf{T}_v = (0, e^v u \cos e^v, -(1/3)\, e^v u \sin e^v)$ We compute

$$\mathbf{T}_u \times \mathbf{T}_v = \begin{vmatrix} \mathbf{i} & \mathbf{j} & \mathbf{k} \\ 2u & \sin e^v & (1/3)\cos e^v \\ 0 & ue^v \cos e^v & (-u/3)e^v \sin e^v \end{vmatrix}$$

$$= ((-u/3)e^v \sin^2 e^v - (u/3)e^v \cos^2 e^v)\mathbf{i} + ((2u^2/3)e^v \sin e^v)\mathbf{j} + (2u^2 e^v \cos e^v)\mathbf{k}$$
$$= (u/3)e^v(-\mathbf{i} + 2u \sin e^v\, \mathbf{j} + 6u \cos e^v\, \mathbf{k}) .$$

To find the tangent plane, we only need to evaluate $-\mathbf{i} + 2u \sin e^v \mathbf{j} + 6u \cos e^v \mathbf{k}$ at the given point (u_0, v_0) . However, we are only given (x_0, y_0, z_0) , so we must be clever. We know that $u \sin e^v = -2$ and $(u/3)\cos e^v = 1$ at the given point, so $2u \sin e^v = -4$ and $6u \cos e^v = 18$. Hence the tangent plane is

$$-(x - 13) - 4(y + 2) + 18(z - 1) = 0 , \text{ or } 18z - 4y - x = 13 .$$

4. (1) A surface is smooth if $\mathbf{T}_u \times \mathbf{T}_v \ne \mathbf{0}$. We calculate $\mathbf{T}_u = 2\mathbf{i} + 2u\mathbf{j} + 0\mathbf{k}$ and $\mathbf{T}_v = 0\mathbf{i} + \mathbf{j} + 2v\mathbf{k}$, so

$$\mathbf{T}_u \times \mathbf{T}_v = \begin{vmatrix} \mathbf{i} & \mathbf{j} & \mathbf{k} \\ 2 & 2u & 0 \\ 0 & 1 & 2v \end{vmatrix} = 4uv\mathbf{i} - 4v\mathbf{j} + 2\mathbf{k} ,$$

which is never $\mathbf{0}$. Therefore, the surface is smooth.

(2) Again, we calculate $\mathbf{T}_u = 2u\mathbf{i} + \mathbf{j} + 2u\mathbf{k}$ and $\mathbf{T}_v = -2v\mathbf{i} + \mathbf{j} + 4\mathbf{k}$, so

$$\mathbf{T}_u \times \mathbf{T}_v = \begin{vmatrix} \mathbf{i} & \mathbf{j} & \mathbf{k} \\ 2u & 1 & 2u \\ -2v & 1 & 4 \end{vmatrix} = (4 - 2u)\mathbf{i} - (8u + 4uv)\mathbf{j} + (2u + 2v)\mathbf{k}$$

If $u = -v = 2$, then $\mathbf{T}_u \times \mathbf{T}_v = \mathbf{0}$. Therefore, the surface is *not* smooth at $(0, 0, -4)$.

7. From the partials of x , y , and z with respect to u , we compute $\mathbf{T}_u = \mathbf{j}$, and similarly, $\mathbf{T}_v = (\cos v)\mathbf{i} - (\sin v)\mathbf{k}$. Then

$$\mathbf{T}_u \times \mathbf{T}_v = \begin{vmatrix} \mathbf{i} & \mathbf{j} & \mathbf{k} \\ 0 & 1 & 0 \\ \cos v & 0 & -\sin v \end{vmatrix} = -(\sin v)\mathbf{i} - (\cos v)\mathbf{k}\,.$$

By our luck, the magnitude of this vector is 1 already, so a unit normal is $\mathbf{n} = -(\sin v)\mathbf{i} - (\cos v)\mathbf{k}$. (Note that another unit normal is $(\sin v)\mathbf{i} + (\cos v)\mathbf{k}$, depending on which cross product ($\mathbf{T}_u \times \mathbf{T}_v$ or $\mathbf{T}_v \times \mathbf{T}_u$) you take).

9. (a) Parametrically, the surface is described by $x = h(y, z)$, $y = y$, $z = z$. In this case, $\mathbf{T}_y = (h_y, 1, 0)$ and $\mathbf{T}_z = (h_z, 0, 1)$, so

$$\mathbf{T}_y \times \mathbf{T}_z = \begin{vmatrix} \mathbf{i} & \mathbf{j} & \mathbf{k} \\ h_y & 1 & 0 \\ h_z & 0 & 1 \end{vmatrix} = (1, -h_y, -h_z)\,.$$

$\mathbf{T}_y \times \mathbf{T}_z$ is normal to the tangent plane, so the tangent plane is $(x - x_0) - h_y(y - y_0) - h_z(z - z_0) = 0$, where h_y and h_z are evaluated at (x_0, y_0, z_0).

10. (b) The surface is parametrized by $(x, y, 3x^2 + 8xy)$, so $\mathbf{T}_x = (1, 0, 6x)$ and $\mathbf{T}_y = (0, 1, 8x)$. At the point $(1, 0, 3)$, $\mathbf{T}_x = (1, 0, 6)$ and $\mathbf{T}_y = (1, 0, 8)$. The normal vector is given by

$$\mathbf{T}_x \times \mathbf{T}_y = \begin{vmatrix} \mathbf{i} & \mathbf{j} & \mathbf{k} \\ 1 & 0 & 6 \\ 1 & 0 & 8 \end{vmatrix} = -2\mathbf{j}\,.$$

So the desired equation is $-2y = 0$ or $y = 0$.

11. (a) When $\theta = 0$, $\mathbf{\Phi}(r, 0)$ is the line segement from $(0, 0, 0)$ to $(1, 0, 0)$. As θ increases, the line segment rotates around the z axis and the segment also moves to the height $z = \theta$. Thus we get the sketch of a helicoid. The sketch in figure 6.4.2 shows a helicoid for θ ranging from 0 to 2π. The helicoid for $0 \le \theta \le 4\pi$ is similar except the graph extends to $z = 4\pi$ and makes an extra revolution around the z axis.

(b) Differentiating the components of $\mathbf{\Phi}(r, \theta)$ gives us $\mathbf{T}_r = (\cos\theta, \sin\theta, 0)$ and $\mathbf{T}_\theta = (-r\sin\theta, r\cos\theta, 1)$. A normal vector to $\mathbf{\Phi}$ is

$$\mathbf{T}_r \times \mathbf{T}_\theta = \begin{vmatrix} \mathbf{i} & \mathbf{j} & \mathbf{k} \\ \cos\theta & \sin\theta & 0 \\ -r\sin\theta & r\cos\theta & 1 \end{vmatrix} = (\sin\theta, -\cos\theta, r)\,.$$

So the unit normal vector is $(\mathbf{T}_r \times \mathbf{T}_\theta)/\|\mathbf{T}_r \times \mathbf{T}_\theta\| = (\sin\theta, -\cos\theta, r)/\sqrt{1 + r^2}$.

(c) Multiply the $\mathbf{T}_r \times \mathbf{T}_\theta$ calculated in part (b) by r to get another normal vector: $(r\sin\theta, -r\cos\theta, r^2) = (y, -x, x^2 + y^2)$. Then the equation of the tangent plane at (x_0, y_0, z_0) is $y_0(x - x_0) - x_0(y - y_0) + (x_0^2 + y_0^2)(z - z_0) = 0$.

15. (a) It is obvious that the image of $\mathbf{\Phi}_1$ is the xy plane (or the uv plane, if you prefer). Since u^3 and v^3 can have any real value, and the third coordinate is 0, the image of $\mathbf{\Phi}_2$ also is the xy plane.

(b) A surface is not smooth if $\| \mathbf{T}_u \times \mathbf{T}_v \| = \mathbf{0}$. For $\mathbf{\Phi}_1$, we have $\mathbf{T}_u = \mathbf{i}$ and $\mathbf{T}_v = \mathbf{j}$, so $\| \mathbf{T}_u \times \mathbf{T}_v \| = \| \mathbf{k} \| = 1$. Therefore, $\mathbf{\Phi}_1$ describes a smooth surface. For $\mathbf{\Phi}_2$, $\mathbf{T}_u = 3u^2\mathbf{i}$ and $\mathbf{T}_v = 3v^2\mathbf{j}$. If $u = 0$ or $v = 0$, then $\mathbf{T}_u \times \mathbf{T}_v = \mathbf{0}$. Hence $\mathbf{\Phi}_2$ is not smooth. (This problem illustrates that "smoothness" depends on parametrization, not necessarily on the image.)

(d) The answer is no. A "smooth" parametrization cannot "round out" the "corners" of a graph (Otherwise it is not a "legal" parametrization of a mapping because its graph would represent that of another mapping)

7.4 Area of a Surface

GOALS

1. Be able to find the surface area of a given surface.

STUDY HINTS

1. *Riemann sum argument.* If you understand the Riemann sum argument, you should be able to derive the surface area formula.

2. *Formulas.* You should know as least one surface area formula. One is

$$A(S) = \int_D \| \mathbf{T}_u \times \mathbf{T}_v \| \, du \, dv .$$

Equation (3) follows immediately from this formula, and (4) is a special case of (3). If you choose to remember equation (3), note that the integrand has the Jacobian matrix of each possible pair chosen from (x, y, z), i.e., (x, y), (y, z), and (x, z).

3. *One-to-one.* The chosen parametrization must be one-to-one; otherwise, the surface may be covered more than once, just as with curves. See exercise 2.

4. *Surface area of a graph.* If $z = f(x, y)$, use the parametrization $x = u$, $y = v$, and $z = f(u, v)$. Then the surface area formula becomes

$$A(S) = \int_D \sqrt{f_x^2 + f_y^2 + 1} \, dA .$$

This is important to know or should be derivable from $\int_D \| \mathbf{T}_u \times \mathbf{T}_v \| \, du \, dv$.

SOLUTIONS TO SELECTED EXERCISES

2. Using formula (3), we compute the following determinants:

$$\frac{\partial(x, y)}{\partial(\theta, \varphi)} = \begin{vmatrix} \partial x/\partial\theta & \partial x/\partial\varphi \\ \partial y/\partial\theta & \partial y/\partial\varphi \end{vmatrix} = \begin{vmatrix} -\sin\theta\sin\varphi & \cos\theta\cos\varphi \\ \cos\theta\sin\varphi & \sin\theta\cos\varphi \end{vmatrix} = -\sin\varphi\cos\varphi\,,$$

$$\frac{\partial(y, z)}{\partial(\theta, \varphi)} = \begin{vmatrix} \cos\theta\sin\varphi & \sin\theta\cos\varphi \\ 0 & -\sin\varphi \end{vmatrix} = -\sin^2\varphi\cos\theta\,, \text{ and}$$

$$\frac{\partial(x, z)}{\partial(\theta, \varphi)} = \begin{vmatrix} -\sin\theta\sin\varphi & \cos\theta\cos\varphi \\ 0 & -\sin\varphi \end{vmatrix} = \sin^2\varphi\sin\theta\,.$$

Squaring the above determinants, adding them, and taking the square root gives us

$$\|\mathbf{T}_\theta \times \mathbf{T}_\varphi\| = \sqrt{\sin^2\varphi\cos^2\varphi + \sin^4\varphi(\cos^2\theta + \sin^2\theta)} = \sqrt{\sin^2\varphi(\cos^2\varphi + \sin^2\varphi)}$$
$$= |\sin\varphi|\,.$$

If we allow φ to vary from $-\pi/2$ to $\pi/2$; then we get

$$\int_0^{2\pi}\int_{-\pi/2}^{\pi/2} |\sin\varphi|\, d\varphi\, d\theta = 2\pi\left[\int_{-\pi/2}^{0} (-\sin\varphi)\, d\varphi + \int_0^{\pi/2} \sin\varphi\, d\varphi\right] =$$
$$2\pi\,[\cos\varphi\Big|_{-\pi/2}^{0} + (-\cos\varphi)\Big|_0^{\pi/2}] = 4\pi\,.$$

The answer is 4π because we change the way φ is measured by varying φ from $-\pi/2$ to $\pi/2$. This parametrization covers the upper hemisphere twice, so the surface area is the same as one complete sphere.

If we vary φ from 0 to 2π, we should get 8π, since the sphere is parametrized twice:

$$\int_0^{2\pi}\int_{\pi}^{2\pi} |\sin\varphi|\, d\varphi\, d\theta = 2\pi\left[\int_0^{\pi} \sin\varphi\, d\varphi + \int_{\pi}^{2\pi} (-\sin\varphi)\, d\varphi\right] =$$
$$2\pi[(-\cos\varphi)\Big|_0^{\pi} + \cos\varphi\Big|_{\pi}^{2\pi} = 2\pi(2 + 2) = 8\pi\,.$$

If we had simply integrated $\sin\varphi$, we would have computed a surface area of 0 in both cases, but this is not possible. Surface areas have to be positive.

5. The area of $\mathbf{\Phi}(D)$ is $\int_D \|\mathbf{T}_u \times \mathbf{T}_v\|\, du\, dv$. Given $\mathbf{\Phi}(u, v) = (u - v, u + v, uv)$, we get

$$\frac{\partial(x, y)}{\partial(u, v)} = \begin{vmatrix} 1 & -1 \\ 1 & 1 \end{vmatrix} = 2\,;\ \frac{\partial(y, z)}{\partial(u, v)} = \begin{vmatrix} 1 & 1 \\ v & u \end{vmatrix} = u - v\,;\ \frac{\partial(x, z)}{\partial(u, v)} = \begin{vmatrix} 1 & -1 \\ v & u \end{vmatrix} = u + v\,.$$

Squaring the determinants, adding them, and then taking the square root gives us the integrand

$$\|\mathbf{T}_u \times \mathbf{T}_v\| = \sqrt{2^2 + (u - v)^2 + (u + v)^2} = \sqrt{4 + 2u^2 + 2v^2} = \sqrt{2}\sqrt{u^2 + v^2 + 2}\,.$$

Since we are integrating over the unit disk, use polar coordinates. Let $u = r\cos\theta$ and $v = r\sin\theta$, $0 \le r \le 1$, $0 \le \theta \le 2\pi$. Then

$$\int_D \sqrt{2}\sqrt{u^2 + v^2 + 2}\, dA = \int_0^1\int_0^{2\pi} \sqrt{2}\sqrt{r^2 + 2}\, r\, d\theta\, dr = 2\sqrt{2}\pi\int_0^1 \sqrt{r^2 + 2}\, r\, dr =$$

$$2\sqrt{2}\pi \cdot \frac{1}{3}(r^2+2)^{3/2}\Big|_0^1 = \frac{2\sqrt{2}\pi}{3}(3^{3/2}-2^{3/2}) = \frac{\pi}{3}(6\sqrt{6}-8).$$

8. We choose the parametrization $x=\sqrt{t^2+1}$, $y=t$, $z=z$, $0\le z\le 1$, $-1\le t\le 1$. To compute the surface area integrand, we first compute

$$\left|\frac{\partial(x,y)}{\partial(t,z)}\right| = 0, \quad \left|\frac{\partial(y,z)}{\partial(t,z)}\right| = 1, \quad \left|\frac{\partial(x,z)}{\partial(t,z)}\right| = \frac{t}{\sqrt{t^2+1}},$$

then,

$$\|\mathbf{T}_t\times\mathbf{T}_z\| = \sqrt{0^2+1^2+\frac{t^2}{t^2+1}}\,dt\,dz = \sqrt{\frac{2t^2+1}{t^2+1}}\,dt\,dz.$$

Thus, the surface area is

$$\int_0^1\int_{-1}^1 \sqrt{\frac{2t^2+1}{t^2+1}}\,dt\,dz = \int_{-1}^1 \sqrt{\frac{2t^2+1}{t^2+1}}\,dt.$$

This integral cannot be done analytically. An alternative parametrization is to use the hyperbolic functions $\sinh t$ and $\cosh t$; the integral only gets nastier.

10. Let $x = u\cos v$, $y = f(u)$, $z = u\sin v$, $a\le u\le b$, $0\le v\le 2\pi$. The reader should verify that

$$\left|\frac{\partial(x,y)}{\partial(u,v)}\right| = -u\,f'(u)\sin v, \quad \left|\frac{\partial(y,z)}{\partial(u,v)}\right| = u\,f'(u)\cos v, \text{ and}$$

$$\left|\frac{\partial(x,z)}{\partial(u,v)}\right| = u.$$

Thus, the surface area is

$$A(S) = \int_D \sqrt{u^2+u^2(f'(u))^2}\,du\,dv.$$

Since the integrand does not depend on v, the v integral can be performed, and we get the desired formula

$$A(S) = 2\pi\int_a^b |u|\sqrt{1+(f'(u))^2}\,du.$$

13. We are interested in the area of the surface $z(x,y)=f(x,y)=1-x-y$, inside $x^2+2y^2\le 1$. First, compute

$$\sqrt{1+f_x^2+f_y^2}\,dx\,dy = \sqrt{3}\,dx\,dy.$$

To compute the surface area, we need to parametrize the disc $z=0$, $x^2+2y^2\le 1$ using polar coordinates: $x=r\cos\theta$, $y=(r/\sqrt{2})\sin\theta$, $0\le r\le 1$, $0\le\theta\le 2\pi$, and the Jacobian is $r/\sqrt{2}$. Our integral then becomes

$$\int_0^{2\pi}\int_0^1 \frac{1}{\sqrt{2}}\,r\cdot\sqrt{3}\,dr\,d\theta = \frac{\pi\sqrt{6}}{2}.$$

17.

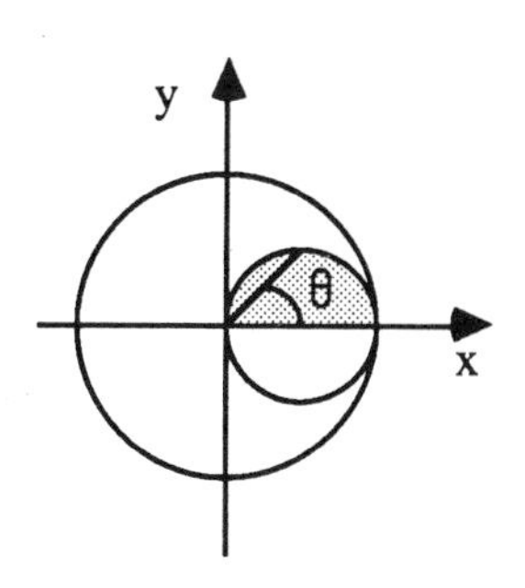

Completing squares, the equation $x^2 + y^2 = x$ becomes $(x^2 - x + 1/4) + y^2 = 1/4$, which is the same as $(x - 1/2)^2 + y^2 = (1/2)^2$. This equation represents a cylinder whose base circle is centered at $(1/2, 0)$ with radius $1/2$, as shown. To find the surface area of S_1, we need to consider where the cylinder "sticks out" of the sphere. Consider the positive octant. The surface area is $\int_D \sqrt{1 + f_x^2 + f_y^2}\,dx\,dy$, where D is half of the base circle (shaded), and $z = f(x, y) = \sqrt{1 - x^2 - y^2}$ is the sphere. Since we will be integrating over a circular region, we can use polar coordinates: $x^2 + y^2 = x$ is the same as $r^2 = r\cos\theta$ or $r = \cos\theta$. From the figure, one can see that D is described by $0 \le r \le \cos\theta$ and $0 \le \theta \le \pi/2$. Also, we compute $f_x = -x/\sqrt{1 - x^2 - y^2}$ and by symmetry, $f_y = -y/\sqrt{1 - x^2 - y^2}$. So $\sqrt{1 + f_x^2 + f_y^2} = 1/\sqrt{1 - x^2 - y^2}$, which becomes $1/\sqrt{1 - r^2}$ in polar coordinates. Remembering that the Jacobian is r and that S_1 consists of four equal surfaces, we get

$$A(S_1) = 4\int_0^{\pi/2}\int_0^{\cos\theta} \frac{r}{\sqrt{1 - r^2}}\,dr\,d\theta = 4\int_0^{\pi/2}\left(-\sqrt{1 - r^2}\,\Big|_{r=0}^{\cos\theta}\right)d\theta$$

$$= 4\int_0^{\pi/2}(1 - \sin\theta)\,d\theta = 4(\theta + \cos\theta)\Big|_0^{\pi/2} = 2\pi - 4\,.$$

By high school geometry, we know that $A(S_2) = 4\pi - (2\pi - 4) = 2\pi + 4$, so $A(S_2)/A(S_1) = (2\pi + 4)/(2\pi - 4)$.

20. First, we are going to figure the volume of the material removed to make the hole in the sphere. Use cylindrical coordinates to describe the hole: $0 \le r \le 1$, $0 \le \theta \le 2\pi$, $-\sqrt{4 - r^2} \le z \le \sqrt{4 - r^2}$. The limits for z were found from the equations of the upper and lower hemispheres: $z = \sqrt{4 - x^2 - y^2}$ and $z = -\sqrt{4 - x^2 - y^2}$ and substituting $r^2 = x^2 + y^2$. Remembering that the Jacobian is r, we get

$$V_{\text{hole}} = \int_0^{2\pi}\int_0^1\int_{-\sqrt{1 - r^2}}^{\sqrt{1 - r^2}} r\,dr\,dz\,d\theta = \int_0^{2\pi}\int_0^1 2r\sqrt{4 - r^2}\,dr\,d\theta$$

$$= \int_0^{2\pi}\left(\frac{-2(4 - r^2)^{3/2}}{3}\,\Big|_{r=0}^{1}\right)d\theta = \frac{2}{3}(8 - 3\sqrt{3})\int_0^{2\pi} d\theta = \frac{4\pi}{3}(8 - 3\sqrt{3})\,.$$

Since the volume of the sphere is $32\pi/3$, the volume of the coupler is $32\pi/3 - (3\pi/3 - 4\pi\sqrt{3}) = 4\pi\sqrt{3}$.

For the surface area, it suffices to calculate the surface area of one "cap" of the hole. In rectangular coordinates, the surface area of $f(x, y) = \sqrt{4 - x^2 - y^2}$ over D, the circle of radius 1, is $\int_D \sqrt{1 + f_x^2 + f_y^2}\,dx\,dy$. We calculate $f_x = -x/\sqrt{4 - x^2 - y^2}$ and

$f_y = -y/\sqrt{4 - x^2 - y^2}$, so $\sqrt{1 + f_x^2 + f_y^2} = 4/\sqrt{4 - x^2 - y^2}$. Changing to polar coordinates, the surface area of one cap is

$$\int_0^{2\pi}\int_0^1 \frac{4r}{\sqrt{4-r^2}}\,dr\,d\theta = \int_0^{2\pi}\left[-4\sqrt{4-r^2}\,\Big|_{r=0}^{1}\right]d\theta = (8 - 4\sqrt{3})\int_0^{2\pi} d\theta = \pi(16 - 8\sqrt{3})\,.$$

Since the surface area of the sphere is $4\pi r^2$, or 16π, and the surface area of the two caps is $2\pi(16 - 8\sqrt{3})$, the outer surface area of the coupler is $16\pi(\sqrt{3} - 1)$.

22. (b) We compute $f_x = y + 1/(y + 1)$ and $f_y = x - x/(y + 1)^2$. Thus

$$\sqrt{1 + f_x^2 + f_y^2}\,dA = \sqrt{1 + \frac{(y^2 + y + 1)^2}{(y + 1)^2} + \frac{(x(y + 1)^2 - x)^2}{(y + 1)^4}}\,dA$$

$$= \frac{1}{(y + 1)^2}\sqrt{(y + 1)^4 + (y^3 + 2y^2 + 2y + 1)^2 + (x(y + 1)^2 - x)^2}\,dA\,,$$

and the surface area is

$$\int_1^4\int_1^2 \frac{1}{(y + 1)^2}\sqrt{(y + 1)^4 + (y^3 + 2y^2 + 2y + 1)^2 + (x(y + 1)^2 - x)^2}\,dy\,dx\,.$$

7.5 Integrals of Scalar Functions Over Surfaces

GOALS

1. Be able to compute the integral of a given scalar function over a given surface.
2. Understand why the integral is defined the way it is and to interpret it physically.

STUDY HINTS

1. *Notation.* In this section, we introduce dS, which stands for $\|\mathbf{T}_u \times \mathbf{T}_v\|\,du\,dv$, which was discussed in section 7.4.
2. *Importance.* This section and the next will be used extensively in chapter 8. In this section, we integrate scalar functions as we did in section 7.1, and in the next section, we will integrate vector-valued functions.
3. *Computation.* The formula for the scalar surface integral of a scalar function f is
$$\int_S f\,dS = \int_D f\,\|\mathbf{T}_u \times \mathbf{T}_v\|\,du\,dv.$$
Here, f is usually given as a function of (x, y, z) and we rewrite it in terms of the parameters u and v by substituting x, y, z as functions of u, v.

4. *Physical interpretation.* (a) If $f = 1$, then we get the surface area. This may help you to remember the formula.
(b) If f is the mass density per unit area at each point of the surface, we get the mass of the surface.

5. *Scalar integral over a graph.* If $z = g(x, y)$, then the formula becomes

$$\int_D f(x, y, g(x, y))\sqrt{1 + \left(\frac{\partial g}{\partial x}\right)^2 + \left(\frac{\partial g}{\partial y}\right)^2}\, dx\, dy .$$

You should remember this or be able to derive it from $\int_D f \, \| \mathbf{T}_u \times \mathbf{T}_v \| \, du\, dv$.

6. *Integrating over a plane.* If S is a plane, we can simplify the integration formula in equation (5):

$$\int_S f\, dS = \int_D \frac{f}{\cos\theta}\, dx\, dy, \quad \text{where} \quad \cos\theta = \mathbf{n} \cdot \mathbf{k}$$

and $\mathbf{n}$ is the *unit* vector normal to the plane. (Review the geometry of the dot product, section 1.2.) We "project" S onto the xy plane to simplify calculations.

SOLUTIONS TO SELECTED EXERCISES

3. Since we're integrating over a hemisphere, it is wise to use spherical coordinates. For the hemispherical surface, we have $\rho = a$, so $x = a\cos\theta\sin\varphi$, $y = a\sin\theta\sin\varphi$, and $z = a\cos\varphi$ for $0 \le \theta \le 2\pi$ and $0 \le \varphi \le \pi/2$. Thus

$$\begin{aligned}\| \mathbf{T}_\theta \times \mathbf{T}_\varphi \| &= \| (-a\sin\theta\sin\varphi\, \mathbf{i} + a\cos\theta\sin\varphi\, \mathbf{j} - 0\mathbf{k}) \times \\ &\qquad (a\cos\theta\cos\varphi\, \mathbf{i} + a\sin\theta\cos\varphi\, \mathbf{j} - a\sin\varphi\, \mathbf{k} \| \\ &= a^2 \sin\varphi .\end{aligned}$$

Then

$$\begin{aligned}\int_S z\, dS &= \iint_D a\cos\varphi \, \| \mathbf{T}_\theta \times \mathbf{T}_\varphi \| \, d\varphi\, d\theta \\ &= \int_0^{2\pi}\int_0^{\pi/2} a\cos\varphi \cdot a^2 \sin\varphi\, d\varphi\, d\theta \\ &= 2\pi a^3 \int_0^{\pi/2} \sin\varphi\cos\varphi\, d\varphi = 2\pi a^3 \left(\frac{\sin^2\varphi}{2}\right)\Bigg|_0^{\pi/2} = \pi a^3 .\end{aligned}$$

7. Parametrize S using polar coordinates: $x = r\cos\theta$, $y = r\sin\theta$, $0 \le r \le 1$, $0 \le \theta \le 2\pi$. Sinced the surface is described by $z = x^2 + y^2$, we substitute in x and y, and get $z = r^2$. Then,

$$\begin{aligned}\mathbf{T}_r \times \mathbf{T}_\theta &= (\cos\theta, \sin\theta, 2r) \times (-r\sin\theta, r\cos\theta, 0) \\ &= (-2r^2\cos\theta, -2r^2\sin\theta, r),\end{aligned}$$

and so

$$dS = \| \mathbf{T}_r \times \mathbf{T}_\theta \| \, dr\, d\theta$$

$$= \sqrt{4r^4 + r^2}\ dr\, d\theta = r\sqrt{4r^2 + 1}\ dr\, d\theta.$$

Calculating $\int_S z\, dS$:

$$\int_S z\, dS = \int_0^{2\pi}\int_0^1 r^2 \cdot r\sqrt{4r^2+1}\ dr\, d\theta$$

$$= 2\pi \int_0^1 r^2 \cdot r\sqrt{4r^2+1}\ dr.$$

This integral can be done using integration by parts (or the tables): let

$$u = r^2,\ \ dv = r\sqrt{4r^2+1}\ dr\,;$$

$$du = 2r\, dr,\ \ v = \frac{1}{12}(4r^2+1)^{3/2}.$$

Then the integral becomes

$$2\pi\left[\frac{r^2}{12}(4r^2+1)^{3/2}\Big|_0^1 - \frac{1}{6}\int_0^1 r\,(4r^2+1)^{3/2} dr\right]$$

$$= 2\pi\left[\frac{5\sqrt{5}}{12} - \frac{1}{6}\cdot\frac{1}{8}\cdot\frac{2}{5}\,(4r^2+1)^{5/2}\Big|_0^1\right]$$

$$= \frac{5\sqrt{5}\,\pi}{12} + \frac{\pi}{60}.$$

8.

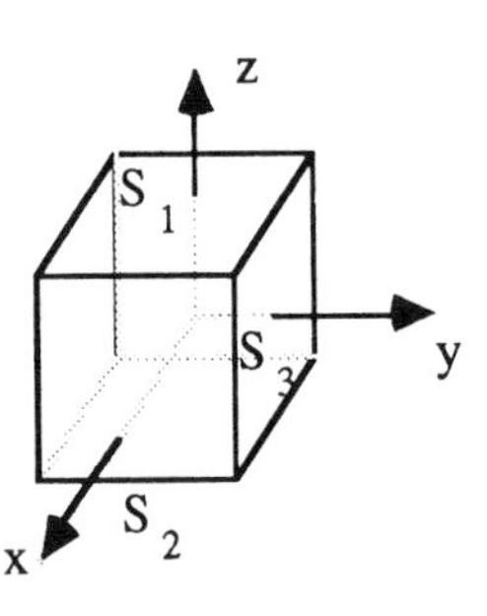

First, integrate over the portion of the cube in the plane $z = 1$, which we will call S_1. We have $-1 \le x \le 1$ and $-1 \le y \le 1$, so $\mathbf{T}_x \times \mathbf{T}_y = \mathbf{i} \times \mathbf{j} = \mathbf{k}$ and $\|\mathbf{T}_x \times \mathbf{T}_y\| = 1$. Then

$$\int_{S_1} z^2\, dS_1 = \iint_D z^2\ \|\mathbf{T}_x \times \mathbf{T}_y\|\ dx\, dy$$

$$= \int_{-1}^1\int_{-1}^1 1\cdot 1\cdot\ dx\, dy = 4.$$

The integral over the portion of the cube in the plane $z = -1$, which we will call S_2, is done in exactly the same way, so

$$\int_{S_2} z^2\, dS_2 = 4.$$

Now, for S_3, which is in the plane $x = 1$. We have D being the square $-1 \le y \le 1$ and $-1 \le z \le 1$, so $\mathbf{T}_y \times \mathbf{T}_z = \mathbf{j} \times \mathbf{k} = \mathbf{i}$ and $\|\mathbf{T}_y \times \mathbf{T}_z\| = 1$. Then

$$\int_{S_3} z^2\, dS_3 = \iint_D z^2\ \|\mathbf{T}_y \times \mathbf{T}_z\|\ dy\, dz = \int_{-1}^1\int_{-1}^1 z^2\, dy\, dz = 2\left(\frac{z^3}{3}\Big|_{-1}^1\right) = \frac{4}{3}.$$

Similarly,

$$\int_{S_4} z^2\, dS_4 = \int_{S_5} z^2\, dS_5 = \int_{S_6} z^2\, dS_6 = \frac{4}{3},$$

where S_4 is the plane $x = -1$, S_5 is the plane $y = 1$, and S_6 is the plane $y = -1$.

Therefore, $\int_S z^2\, dS$ is the sum of the integrals over the six surfaces, which is $4 + 4 + 4/3 + 4/3 + 4/3 + 4/3 = 40/3$.

11. (a) The equation of the sphere is $x^2 + y^2 + z^2 = R^2$. One can substitute x for y, y for z, and z for x, or any other permutation, and still get the same equation. So the three integrals ought to be equal. (This is what "by symmetry" usually means.) Geometrically, a sphere "looks the same" no matter how you look at it.

(b) Using part (a), we have

$$\int_S (x^2 + y^2 + z^2)\, dS = \int_S x^2\, dS + \int_S y^2\, dS + \int_S z^2\, dS = 3\int_S x^2\, dS .$$

Substitute $x^2 + y^2 + z^2 = R^2$ and recall that $\int_S dS$ is the surface area of S to get

$$\int_S x^2\, dS = \frac{1}{3}\int_S (x^2 + y^2 + z^2) dS = \frac{1}{3}\int_S R^2\, dS = \frac{R^2}{3} \cdot 4\pi R^2 = \frac{4\pi}{3} R^4 .$$

(c) Due to the symmetry of the sphere, if we integrate $x^2 + y^2$ over the entire sphere, we should get twice the mass desired in exercise 10. So the desired mass is

$$\frac{1}{2}\int_S (x^2 + y^2) dS = \frac{1}{2}(2)\int_S x^2\, dS = \frac{4\pi}{3} R^4 .$$

14. By exercise 12, the average z-coordinate is

$$\frac{1}{A(S)}\int_S z\, dS .$$

In this case, $A(S)$, the surface area of a hemisphere of radius R, is $2\pi R^2$. By exercise 3, $\int_S z\, dS$ is πR^3, so $\bar{z} = \pi R^3/2\pi R^2 = R/2$. Since there is as much of the sphere on one side of z axis than on the other (by symmetry), $\bar{x} = \bar{y} = 0$.

17. (a)

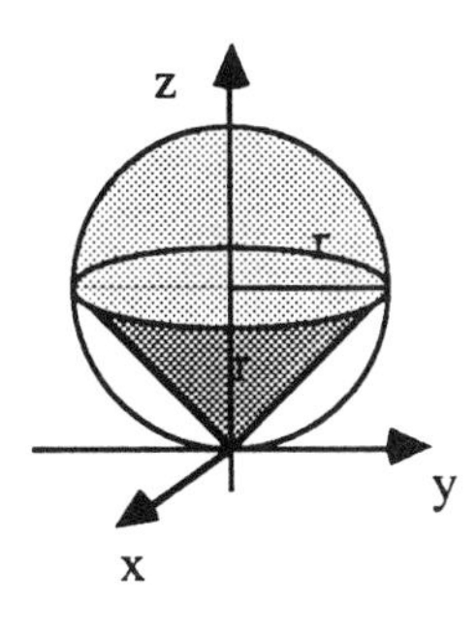

The equation of the sphere is $x^2 + y^2 + z^2 = 2rz$, $r > 0$. Upon completing the squares, we see that it is equal to $x^2 + y^2 + (z - r)^2 = r^2$, which is a sphere of radius r centered at $(0, 0, r)$. The tip of the cone $z^2 = x^2 + y^2$ intersects (by design) the sphere at the origin. To find the other intersections, note that $x^2 + y^2 + (z - r)^2 = r^2$ and $x^2 + y^2 = z^2$ implies $r^2 - (z - r)^2 = z^2$ or $z = r$. Parametrize the cone as follows:

$x = \rho \cos\theta$, $y = \rho \sin\theta$, $z = \rho$ with $0 \le \rho \le r$ and $0 \le \theta \le 2\pi$. You should verify that $\| \mathbf{T}_\rho \times \mathbf{T}_\theta \| = \sqrt{2}\,\rho$. Thus the area of the cone that is inside the sphere is

$$\int_0^{2\pi}\int_0^r \sqrt{2}\rho\, d\rho\, d\theta = 2\pi\sqrt{2}\,\frac{\rho^2}{2}\Big|_0^r = \pi\sqrt{2}r^2 .$$

(b) By "area of the portion of the sphere inside the cone," the authors mean, we think, the area of the piece of the sphere that is the "ice cream" part of this configuaration. Well, this is trivial; it is simply the area of the hemisphere of radius r, or $2\pi r^2$.

7.6 Surface Integrals of Vector Functions

GOALS

1. Be able to compute a surface integral of a vector function.
2. Understand its derivation and physical interpretation.

STUDY HINTS

1. *Notation.* The symbol $\mathbf{n}(\mathbf{\Phi}(u_0, v_0))$ is used to describe the unit vector which is normal to (a parametrized surface) $\mathbf{\Phi}$ at (u_0, v_0) ; note that $\mathbf{\Phi}(u_0, v_0)$ is the base point of $\mathbf{n}$.
2. *Orientation.* As with line integrals, the orientation is important. The sign of the integral changes with the opposite orientation.
3. *Parametrization.* As with line integrals, we can reparametrize a surface. As long as the orientation is retained, the value of the integral is unchanged.
4. *Definition.* The surface integral of a vector field $\mathbf{F}$ over a parametrized surface $\mathbf{\Phi}$ is
$$\int_{\mathbf{\Phi}} \mathbf{F} \cdot d\mathbf{S} = \int_D \mathbf{F} \cdot (\mathbf{T}_u \times \mathbf{T}_v)\, du\, dv .$$
The integral of a vector field is a scalar.
5. *Generalizations.* Notice that scalar integrals (sections 7.1 and 7.5) do not depend on orientation. The sign will change only in the integration of a vector field (sections 7.2 and 7.6) if the orientation has been reversed. Finally, note that scalar integrals involve the length of a vector and the integral of vector fields involve the dot product.
6. *Reduction to scalar integrals.* If we know the unit vector to the surface $\mathbf{\Phi}$, then the surface integral reduces to $\int_S (\mathbf{F} \cdot \mathbf{n})\, dS$. Letting $f = \mathbf{F} \cdot \mathbf{n}$, we get the scalar integral of section 7.5. If the orientation switches, $\mathbf{n}$ changes sign and so does this f .
7. *Surface integral over a graph.* If $z = f(x, y)$ and $\mathbf{F} = (F_1, F_2, F_3)$, we know that $\mathbf{n} = (-\partial f/\partial x, -\partial f/\partial y, 1)$, so the surface integral becomes
$$\int_D \left[F_1\left(-\frac{\partial f}{\partial x}\right) + F_2\left(-\frac{\partial f}{\partial y}\right) + F_3 \right] dx\, dy .$$
8. *Physical interpretation.* The surface integral over $\mathbf{\Phi}$ gives the rate of flow of a fluid across the surface $\mathbf{\Phi}$. This is known as flux.
9. *Unit normal to unit sphere.* It is useful to know that for the *unit sphere,* $\mathbf{n} = \mathbf{r} = (x, y, z)$.

10. *Good example.* Example 6 shows three ways of computing the same integral.

SOLUTIONS TO SELECTED EXERCISES

1. As in example 4, the heat flux across the surface S is $\int_S -\nabla T \cdot d\mathbf{S}$. Parametrize the surface $x^2 + z^2 = 2$. Let $x = \sqrt{2}\cos\theta$, $z = \sqrt{2}\sin\theta$ with $0 \le \theta \le 2\pi$. So the surface can be parametrized by $S(\theta, y) = (\sqrt{2}\cos\theta, y, \sqrt{2}\sin\theta)$, $0 \le \theta \le 2\pi$ and $0 \le y \le 2$. Then the outward pointing normal to S is $\mathbf{T}_\theta \times \mathbf{T}_y = \sqrt{2}\cos\theta\,\mathbf{i} + \sqrt{2}\sin\theta\,\mathbf{k}$. Given $T(x, y, z) = 3x^2 + 3z^2$, we compute $-\nabla T = -6(x, 0, z) = -6(\sqrt{2}\cos\theta, 0, \sqrt{2}\sin\theta)$. Thus

$$\int_S -\nabla T \cdot d\mathbf{S} = \iint_D -\nabla T \cdot (\mathbf{T}_\theta \times \mathbf{T}_y)\, dy\, d\theta$$

$$= -6\int_0^{2\pi}\int_0^2 [(\sqrt{2}\cos\theta)^2 + (\sqrt{2}\sin\theta)^2]\, dy\, d\theta$$

$$= -12\int_0^{2\pi}\int_0^2 dy\, d\theta = -48\pi .$$

If the other cross product (hence the opposite orientation) is used, the answer would be $+48\pi$.

6. First, we compute

$$\nabla \times \mathbf{F} = \begin{vmatrix} \mathbf{i} & \mathbf{j} & \mathbf{k} \\ \partial/\partial x & \partial/\partial y & \partial/\partial z \\ x^2 + y - 4 & 3xy & 2xz + z^2 \end{vmatrix} = -2z\mathbf{j} + (3y - 1)\mathbf{k} .$$

Use spherical coordinates to parametrize S: $x = 4\cos\theta\sin\varphi$, $y = 4\sin\theta\sin\varphi$, $z = 4\cos\varphi$ with $0 \le \varphi \le \pi/2$ and $0 \le \theta \le 2\pi$. Then $\mathbf{T}_\theta \times \mathbf{T}_\varphi = 16(-\sin^2\varphi\cos\theta\mathbf{i} - \sin^2\varphi\sin\theta\mathbf{j} - \sin\varphi\cos\varphi\mathbf{k})$. Thus

$$\int_S (\nabla \times \mathbf{F}) \cdot d\mathbf{S} = \int_S (0, -2z, 3y - 1) \cdot (\mathbf{T}_\theta \times \mathbf{T}_\varphi)\, d\theta\, d\varphi =$$

$$-16\int_S [(0, -8\cos\varphi, 12\sin\theta\sin\varphi - 1) \cdot (\sin^2\varphi\cos\theta, \sin^2\varphi\sin\theta, \sin\varphi\cos\varphi)]\, d\theta\, d\varphi =$$

$$-16\int_0^{2\pi}\int_0^{\pi/2} (4\sin\theta\sin^2\varphi\cos\varphi - \sin\varphi\cos\varphi)\, d\varphi\, d\theta =$$

$$-16\int_0^{2\pi}\left[\left(\frac{4}{3}\sin^3\varphi\sin\theta - \frac{1}{2}\sin^2\varphi\right)\Big|_0^{\pi/2}\right] d\theta = -16\int_0^{2\pi}\left(\frac{4}{3}\sin\theta - \frac{1}{2}\right) d\theta =$$

$$-16\left[\frac{-4}{3}\cos\theta - \frac{\theta}{2}\right]\Big|_0^{2\pi} = 16\pi .$$

8. (a) The wall lies under the circle $z = 4R^2$, $x^2 + (y - R)^2 = R^2$ and above the mountain $x^2 + y^2 + z = 4R^2$. From the top view, we may parametrize the circle by

$$x = R\cos\theta,\ y - R = R\sin\theta,\ 0 \le \theta \le 2\pi.$$

Then the mountain becomes

$$z = 4R^2 - (x^2 + y^2) = 4R^2 - [(R\cos\theta)^2 + (R + R\sin\theta)^2]$$
$$= 4R^2 - [2R^2 + 2R\sin\theta] = 2R^2 - 2R\sin\theta.$$

To find the surface area of the "cylindrical" wall of the restaurant, we parametrize the wall by

$$x = R\cos\theta,\ y = R + R\sin\theta,\ z = z,$$
$$0 \le \theta \le 2\pi,\ 2R^2 - 2R\sin\theta \le z \le 4R^2,$$

and the surface area becomes

$$\int_0^{2\pi}\int_{2R^2 - 2R\sin\theta}^{4R^2} \|\mathbf{T}_\theta \times \mathbf{T}_z\|\ dz\,d\theta.$$

The reader should verify that $\|\mathbf{T}_\theta \times \mathbf{T}_z\| = R$. So the integral becomes

$$\int_0^{2\pi}\int_{2R^2 - 2R\sin\theta}^{4R^2} R\,dz\,d\theta = \int_0^{2\pi} R(4R^2 - 2R^2 + 2R\sin\theta)\,d\theta = 4\pi R^3.$$

(b) Parametrize the restaurant by

$$x = r\cos\theta,\ y = r + r\sin\theta,\ z = z,$$

where $0 \le r \le R$, $0 \le \theta \le 2\pi$, $4R^2 - (2r^2 + 2r\sin\theta) \le z \le 4R^2$. The Jacobian is

$$\frac{\partial(x, y, z)}{\partial(r, \theta, z)} = \begin{vmatrix} \cos\theta & -r\sin\theta & 0 \\ 1 + \sin\theta & r\cos\theta & 0 \\ 0 & 0 & 1 \end{vmatrix} = r + r\sin\theta.$$

So the volume of the restaurant becomes

$$\int_0^{2\pi}\int_0^R\int_{4R^2 - (2r^2 + 2r\sin\theta)}^{4R^2} (r + r\sin\theta)\,dz\,dr\,d\theta$$
$$= \int_0^{2\pi}\int_0^R (r + r\sin\theta)(4R^2 - 4R^2 + 2r^2 + 2r\sin\theta)\,dr\,d\theta$$
$$= \int_0^{2\pi}\int_0^R (2r^3 + 2r^2\sin\theta +$$
$$2r^3\sin\theta + 2r^2\sin^2\theta)dr\,d\theta$$
$$= \int_0^{2\pi}\left(\frac{R^4}{2} + \frac{2}{3}R^3\sin\theta + r^3\sin\theta + \frac{R^4}{2}\sin\theta + \frac{2R^3}{3}\sin^2\theta\right)d\theta.$$

Since $\int_0^{2\pi}\sin\theta\,d\theta = 0$ and $\int_0^{2\pi}\sin^2\theta\,d\theta = \pi$, the volume becomes

$$\pi R^4 + \frac{2\pi}{3}R^3.$$

To determine whether the volume exceeds $\pi R^4/2$, we solve the inequality $V \geq \pi R^4/2$. Keeping in mind that $R > 0$, we see that the restaurant will be a profit maker for all $R > 0$.

(c) The heat flux is $\int_S \mathbf{V} \cdot d\mathbf{S}$, where $\mathbf{V} = -k(6x, 2y - 2R, 32z)$. The roof of the restaurant can be parametrized by $x = r\cos\theta$, $y = r\sin\theta + R$, $z = 4R^2$ for $0 \leq r \leq R$ and $0 \leq \theta \leq 2\pi$. We have $\mathbf{T}_r = (\cos\theta)\mathbf{i} + (\sin\theta)\mathbf{j}$ and $\mathbf{T}_\theta = (-r\sin\theta)\mathbf{i} + (r\cos\theta)\mathbf{j}$, so $\mathbf{T}_r \times \mathbf{T}_\theta = r\mathbf{k}$. Therefore, the heat flux through the roof is

$$= -k\int_S (6x, 2y - 2R, 32z) \cdot (0, 0, r)\, dr\, d\theta = -k\int_0^{2\pi}\int_0^R 32zr\, dr\, d\theta$$

$$= -k\int_0^{2\pi}\int_0^R 64Rr\, dr\, d\theta = -k\int_0^{2\pi} \left(32Rr^2\Big|_{r=0}^{R}\right) d\theta$$

$$= -k(32R^3)(2\pi) = -64\pi R^3 k .$$

The side which makes contact with the mountain can be parametrized by $x = r\cos\theta$, $y = r\sin\theta + R$, $z = 4R^2 - (2r^2 + 2r\sin\theta)$ for $0 \leq r \leq R$ and $0 \leq \theta \leq 2\pi$. We have $\mathbf{T}_r = (\cos\theta)\mathbf{i} + (\sin\theta)\mathbf{j} + (-2\sin\theta)\mathbf{k}$ and $\mathbf{T}_\theta = (-r\sin\theta)\mathbf{i} + (r\cos\theta)\mathbf{j} + (-2r\cos\theta)\mathbf{k}$, so $\mathbf{T}_r \times \mathbf{T}_\theta = (-4r^2\cos\theta)\mathbf{i} + (2r + 4r^2\sin\theta)\mathbf{j} + r\mathbf{k}$. Therefore, the heat flux through the mountain side is

$$-k\int_S (6x, 2y - 2R, 32z) \cdot (-4r^2\cos\theta, 2r + 4r^2\sin\theta, r)\, dr\, d\theta$$

$$= -k\int_0^R\int_0^{2\pi} (-24r^3\cos^2\theta + 4r^2\sin\theta + 8r^3\sin^2\theta + 128R^2r - 64r^3 - 64r^2\sin\theta)\, d\theta\, dr$$

$$= -k\int_0^R (-24r^3\pi + 8r^3\pi + 256R^2r\pi - 128r^3\pi)dr = -92R^4\pi k .$$

Finally, the curved glass wall can be parametrized as in part (a). We have $\mathbf{T}_\theta \times \mathbf{T}_z = (R\cos\theta)\mathbf{i} + (R\sin\theta)\mathbf{j}$. The heat flux through the wall is

$$-k\int_S (6x, 2y - 2R, 32z) \cdot (R\cos\theta, R\sin\theta, 0)\, dz\, d\theta$$

$$= -k\int_0^{2\pi}\int_{2R^2 - 2R\sin\theta}^{4R^2} (6R^2\cos^2\theta + 2R^2\sin^2\theta)\, dz\, d\theta$$

$$= -k\int_0^{2\pi} (4R^2 + 2R^2\cos^2\theta)(2R^2 - 2R\sin\theta)\, d\theta$$

$$= -20k\pi R^4 .$$

Adding these reults together, we find that the total flux is $-112k\pi R^4$.

11. The surface S is the unit sphere so S can be parametrized by $x = \cos\theta\sin\varphi$, $y = \sin\theta\sin\varphi$ and $z = \cos\varphi$ for $0 \leq \theta \leq 2\pi$ and $0 \leq \varphi \leq \pi$. Differentiate each

component with respect to θ to get $\mathbf{T}_\theta = (-\sin\theta \sin\varphi, \cos\theta \sin\varphi, 0)$. Similarly, $\mathbf{T}_\varphi = (\cos\theta \cos\varphi, \sin\theta \cos\varphi, -\sin\varphi)$. The normal vector for S is $\mathbf{T}_\varphi \times \mathbf{T}_\theta =$ $(\sin^2\varphi \cos\theta, \sin^2\varphi \sin\theta, \sin\varphi \cos\varphi)$. Factor out $\sin\varphi$ to get $(\sin\varphi)(x, y, z) =$ $\mathbf{n} = (\sin\varphi)\,\mathbf{r}$. Suppose $\mathbf{F} = (F_r, F_\theta, F_\varphi)$, then $\mathbf{F} \cdot \mathbf{n} = \mathbf{F} \cdot (\sin\varphi\ \mathbf{r}) = F_r \sin\varphi$. Here, we make use of the fact that $(\mathbf{e}_r, \mathbf{e}_\theta, \mathbf{e}_\varphi)$ form an orthonormal basis (see section 1.4). Hence

$$\int_S \mathbf{F} \cdot d\mathbf{S} = \int_0^{2\pi}\int_0^{\pi} F_r \sin\varphi \, d\varphi \, d\theta .$$

The corresponding formula for real-valued functions is

$$\int_S f \, dS = \int_0^{2\pi}\int_0^{\pi} f \sin\varphi \, d\varphi \, d\theta.$$

14. A parametrzation of the surface is $\mathbf{\Phi}(u, v) = (u, v, 0)$, so $\|\mathbf{T}_u \times \mathbf{T}_v\| = \|\mathbf{i} \times \mathbf{j}\| =$ $\|\mathbf{k}\| = 1$. The scalar integral is

$$\int_S f(x, y, z)\, dS = \iint_D f(u, v, 0)\ \|\mathbf{T}_u \times \mathbf{T}_v\| \, du \, dv .$$

By substituting (x, y) for (u, v), we get $\iint_D f(x, y, 0)\, dx\, dy$. For a vector field $\mathbf{F} =$ (F_x, F_y, F_z) (NOTE: The subscripts here do *not* denote partials!),

$$\int_S \mathbf{F} \cdot d\mathbf{S} = \iint_D \mathbf{F} \cdot \mathbf{k} \, dx \, dy = \iint_D F_z \, dx \, dy .$$

That is, only the z component of F matters.

16. (a) Parametrize the cone as follows: $x = r\cos\theta$, $y = r\sin\theta$, $z = (x^2 + y^2)^{1/2} = r$, $0 \le r \le 1$, $0 \le \theta \le 2\pi$. Then $\mathbf{T}_\theta \times \mathbf{T}_r = (-r\sin\theta, r\cos\theta, 0) \times (\cos\theta, \sin\theta, 1)$ $= (r\cos\theta, r\sin\theta, -r)$. The flux is the surface integral of $\mathbf{F}$ over S, which is

$$\int_0^{2\pi}\int_0^{1} (0, 0, -1) \cdot (r\cos\theta, r\sin\theta, -r)\, dr\, d\theta = \int_0^{2\pi}\int_0^{1} r \, dr \, d\theta = \pi .$$

For the hot shots, note that the same amount of rain falling through the cone must also go through the disk $x^2 + y^2 \le 1$, $z = 0$. The problem then could be done in your head.

(b) Using the same parametrization as in part (a), the total flux through the cone is

$$\int_0^{2\pi}\int_0^{1}\left(\frac{-\sqrt{2}}{2}, 0, \frac{-\sqrt{2}}{2}\right) \cdot (r\cos\theta, r\sin\theta, -r)\, dr\, d\theta$$

$$= \int_0^{2\pi}\int_0^{1} \frac{\sqrt{2}}{2}(-r\cos\theta + r)\, dr\, d\theta$$

$$= \frac{\sqrt{2}}{2}\int_0^{2\pi} \frac{1}{2}(1 - \cos\theta)\, d\theta = \frac{\pi}{2}\sqrt{2} .$$

7.R Review Exercises for Chapter 7

1. (b) First, we calculate $\boldsymbol{\sigma}'(t) = (-\sin t, \cos t, 1)$, so $\|\boldsymbol{\sigma}'(t)\| = \sqrt{2}$. Then the path integral is

$$\int_\sigma f\, ds = \int_0^{2\pi} xyz\, \|\boldsymbol{\sigma}'(t)\|\, dt = \int_0^{2\pi} \sqrt{2}(\cos t)(\sin t)t\, dt .$$

Use integration by parts with $u = t$, $dv = \cos t \sin t\, dt$, $du = dt$, and $v = (1/2)\sin^2 t$. The integral becomes

$$\sqrt{2}\left(\frac{t}{2} \sin^2 t \bigg|_0^{2\pi} - \int_0^{2\pi} \frac{1}{2} \sin^2 t\, dt \right) = -\frac{\sqrt{2}}{2} \int_0^{2\pi} \frac{1 - \cos 2t}{2}\, dt = \frac{-\pi\sqrt{2}}{2} .$$

2. (b) As in exercise 1(b), we have $ds = \|\boldsymbol{\sigma}'(t)\|\, dt = \sqrt{2}\, dt$. Then, the path integral is

$$\int_\sigma f\, ds = \int_0^{2\pi} (\sin t + \cos^2 t)\sqrt{2}\, dt = \sqrt{2}\int_0^{2\pi} \left(\sin t + \frac{1 + \cos 2t}{2} \right) dt$$

$$= \sqrt{2}\left(-\cos t + \frac{t}{2} + \frac{\sin 2t}{4} \right) \bigg|_0^{2\pi} = \sqrt{2}\pi .$$

3. (a) Compute the line integral over each segment of C and add them together. Be careful with the orientation. Recall from chapter 1 that the equation $\mathbf{l}(t) = (x, y, z) = (1, 0, 0) + t(-1, 1, 0)$, $0 \le t \le 1$ satisfies the condition $\mathbf{l}(0) = (1, 0, 0)$ and $\mathbf{l}(1) = (0, 1, 0)$. On this segment, $x = 1 - t$, $y = t$, and $z = 0$. Also, $dx = -dt$, $dy = dt$, and $dz = 0$. Substitute these values to get

$$\int_0^1 [\sin \pi(1 - t)dt - (\cos \pi t)(0)] = \int_0^1 \sin \pi(1 - t)\, dt$$

$$= \frac{1}{\pi} \cos \pi(1 - t) \bigg|_0^1 = \frac{1}{\pi}(1 - (-1)) = \frac{2}{\pi} .$$

The next segment can be parametrized by $\mathbf{l}(t) = (x, y, z) = (0, 1, 0) + t(0, -1, 1) = (0, 1 - t, t)$ for $0 \le t \le 1$. Also, $dx = 0$, $dy = -dt$, and $dz = dt$, so we get

$$\int_0^1 [\sin \pi(0)\, (-dt) - \cos \pi(1 - t)\, dt] = \frac{1}{\pi} \sin \pi(1 - t) \bigg|_0^1 = 0 .$$

Fianlly, the last segment can be parametrized by $\mathbf{l}(t) = (x, y, z) = (0, 0, 1) + t(1, 0, -1) = (t, 0, 1 - t)$ for $0 \le t \le 1$. Also, $dx = dt$, $dy = 0$, and $dz = -dt$, so we get

$$\int_0^1 [\sin \pi t\, (0) - \cos \pi(0)\, (-dt)] = \int_0^1 1\, dt = 1 .$$

Therefore, the line integral around the triangle is $2/\pi + 0 + 1 = 2/\pi + 1$.

5. We start at $(0, 0)$ and compute the line integral over each side of the square. For the line segment from $(0, 0)$ to $(a, 0)$, $y = 0$, $dy = 0$, and $0 \le x \le a$. So

$$\int \mathbf{F} \cdot d\mathbf{s} = \int_0^a [(x^2 - 0^2)dx + 2x(0) \cdot (0)] = \int_0^a x^2\, dx = \frac{a^3}{3} .$$

For the line segment from $(a, 0)$ to (a, a), $x = a$, $dx = 0$, and $0 \le y \le a$. So

$$\int \mathbf{F} \cdot d\mathbf{s} = \int_0^a [(a^2 - y^2)(0) + 2(a)y\, dy] = \int_0^a 2ay\, dy = ay^2 \bigg|_0^a = a^3 .$$

For the line segment from (a, a) to (0, a) , $y = a$, $dy = 0$, and x goes from a to 0 . Thus

$$\int \mathbf{F} \bullet d\mathbf{s} = \int_a^0 (x^2 - a^2)dx = -\int_0^a (x^2 - a^2)dx = -\left(\frac{a^3}{3} - a^3\right) = \frac{2a^3}{3} .$$

Finally, for the line segment from (0, a) back to (0, 0) , $x = 0$, $dx = 0$, and y goes from a to 0 . Thus

$$\int \mathbf{F} \bullet d\mathbf{s} = 0 .$$

By addition, we find that the line integral around the square is $a^3/3 + a^3 + 2a^3/3 + 0 = 2a^3$.

7. (b) First, complete the squares: $(2x^2 - 8x + 8) + y^2 + z^2 = 1 + 8$ or $2(x-2)^2 + y^2 + z^2 = 3^2$. This has the form $X^2 + y^2 + z^2 = \rho^2$, so use spherical coordinates: For the x parametrization, let $X = \sqrt{2}(x-2) = 3 \cos\theta \sin\varphi$, so $x = 2 + (3 \cos\theta \sin\varphi)/\sqrt{2}$. Thus, the parametrization is

$$x = 2 + (3 \cos\theta \sin\varphi)/\sqrt{2}$$
$$y = 3 \sin\theta \sin\varphi$$
$$z = 3 \cos\varphi ,$$

with $0 \le \theta \le 2\pi$ and $0 \le \varphi \le \pi$. This is a general strategy in attacking many parametrization problems: changing cartesian coordinates into either cylindrical or spherical coordinates by completeing the squares.

10. The surface area of the graph lying over D is $\iint_D \| \mathbf{T}_x \times \mathbf{T}_y \| \, dx\, dy$ and the area of D is $\iint_D dx\, dy$. The "parametrization" of the graph is $x = x, y = y, z = f(x, y)$. So $\mathbf{T}_x = \mathbf{i} + (\partial f/\partial x)\mathbf{k}$ and $\mathbf{T}_y = \mathbf{j} + (\partial f/\partial y)\mathbf{k}$. Therefore, $\mathbf{T}_x \times \mathbf{T}_y = (\partial f/\partial x)\mathbf{i} + (\partial f/\partial y)\mathbf{j} + \mathbf{k}$ and $\| \mathbf{T}_x \times \mathbf{T}_y \| = [(\partial f/\partial x)^2 + (\partial f/\partial y)^2 + 1]^{1/2}$. Since $(\partial f/\partial x)^2 + (\partial f/\partial y)^2 = c$, $\| \mathbf{T}_x \times \mathbf{T}_y \| = \sqrt{1 + c}$. Returning to the original formula, we have $\iint_D \| \mathbf{T}_x \times \mathbf{T}_y \| \, dx\, dy = \iint_D \sqrt{1+c} \, dx\, dy$. Since c is a constant, we factor the constant from the integral to get $\sqrt{1+c} \iint_D dx\, dy = \sqrt{1+c} \bullet (\text{Area of D})$.

12. (b) Use cylindrical coordinates. Let $x = r \cos\theta$, $y = r \sin\theta$. In addition, $z = x = r \cos\theta$, and the intervals are $0 \le r \le 1$, $0 \le \theta \le 2\pi$ since we want to be inside the cylinder $x^2 + y^2 = 1$. We calculate $\mathbf{T}_r \times \mathbf{T}_\theta = (\cos\theta\, \mathbf{i} + \sin\theta\, \mathbf{j} + \cos\theta\, \mathbf{k}) \times (-r \sin\theta\, \mathbf{i} + r \cos\theta\, \mathbf{j} + r \sin\theta\, \mathbf{k}) = (-r\mathbf{i} + r\mathbf{k}) = r(-\mathbf{i} + \mathbf{k})$, so $\| \mathbf{T}_r \times \mathbf{T}_\theta \| = \sqrt{2}\, r$. Therefore,

$$\int x^2 \, dS = \int_0^{2\pi} \int_0^1 (r \cos\theta)^2 \sqrt{2} r \, dr\, d\theta = \left(\sqrt{2} \int_0^1 r^3 \, dr\right)\left(\int_0^{2\pi} \cos^2\theta \, d\theta\right)$$

$$= \left(\frac{\sqrt{2}}{4}\right)\left(\int_0^{2\pi} \frac{1+\cos 2\theta}{2}\, d\theta\right) = \frac{\sqrt{2}}{4}\left(\frac{\theta}{2} + \frac{\sin 2\theta}{4}\right)\Big|_0^{2\pi} = \frac{\sqrt{2}}{4}\pi .$$

15. 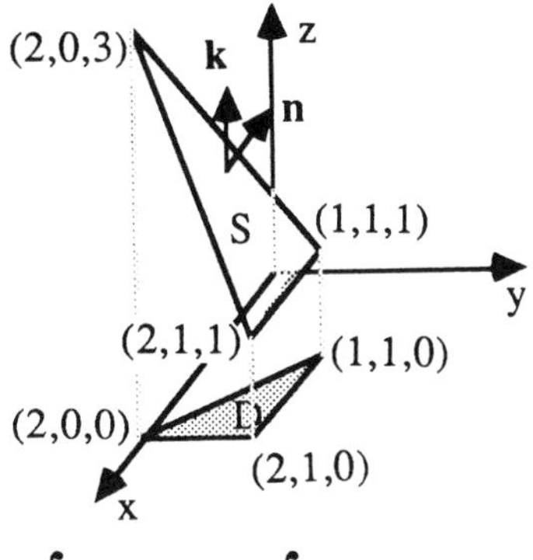

We want to compute $\int_S x\, dS$, where S is the triangle with vertices (1, 1, 1), (2, 1, 1) and (2, 0, 3). First, we need to find the normal to the triangle: two vectors on the triangle are (1, 0, 0) and (0, 1, −2) (found by subtracting the coordinates of the vertices). Take their cross product and normalize it, we get the unit normal $\mathbf{n} = (0, 2, 1)/\sqrt{5}$, so $\cos\theta = \mathbf{n} \cdot \mathbf{k} = 1/\sqrt{5}$. Next, the projection of S onto the xy-plane can be described by $-y + 2 \le x \le 2,\ 0 \le y \le 1$, as shown. Thus,

$$\int_S x\, dS = \sqrt{5}\int_D x\, dx\, dy = \sqrt{5}\int_0^1\int_{-y+2}^{2} x\, dx\, dy$$

$$= \frac{\sqrt{5}}{2}\int_0^1 4 - (2-y)^2\, dy$$

$$= \frac{\sqrt{5}}{2}\left[4 + \frac{(2-y)^3}{3}\Big|_0^1\right] = \frac{5\sqrt{5}}{6} .$$

19. Here, we have $\mathbf{F} = e^t\mathbf{i} + t\mathbf{j} + t^2\mathbf{k}$ and $d\mathbf{s} = e^t\mathbf{i} + \mathbf{j} + 2t\mathbf{k}$. Thus

$$\int_C \mathbf{F} \cdot d\mathbf{s} = \int_0^1 (e^t \cdot e^t + t \cdot 1 + t^2 \cdot 2t)\, dt = \int_0^1 (e^{2t} + t + 2t^3)dt$$

$$= \left(\frac{e^{2t}}{2} + \frac{t^2}{2} + \frac{t^4}{2}\right)\Big|_0^1 = \frac{1}{2}(e^2 + 1) .$$

24. Our surface is $z^2 = 1 - x^2 - y^2$. We want the part *above* the xy plane, so we want $z \ge 0$, or $f(x, y) = z = (1 - x^2 - y^2)^{1/2}$. For the surface area of a graph over D , the formula is

$$A = \int_D \sqrt{1 + f_x^2 + f_y^2}\, dx\, dy .$$

In this case,

$$f_x = \frac{-1}{2} \cdot 2x\,(1 - x^2 - y^2)^{-1/2} = \frac{-x}{(1 - x^2 - y^2)^{1/2}} \quad \text{and}$$

$$f_y = \frac{-y}{(1 - x^2 - y^2)^{1/2}} \quad \text{(by symmetry)} .$$

So the integrand is

$$\sqrt{1 + \frac{x^2 + y^2}{1 - x^2 - y^2}} = \frac{1}{(1 - x^2 - y^2)^{1/2}} .$$

The area we want is

$$\int_{-a}^{a}\int_{-a}^{a}\frac{1}{(1-x^2-y^2)^{1/2}}\,dy\,dx\,.$$

We can evaluate one of the integrals. Let $u^2 = (1-x^2)^{1/2}$. Since u is independent of y, the integral in y becomes

$$\int_{-a}^{a}\frac{1}{(u^2-y^2)^{1/2}}\,dy = \frac{1}{u}\int_{-a}^{a}\frac{1}{\sqrt{1-(y/u)^2}}\,dy\,.$$

Substituting $v = y/u$ with $dy = u\,dv$, we get

$$\int_{-a/u}^{a/u}\frac{dv}{\sqrt{1-v^2}} = \sin^{-1}(v)\Big|_{-a/u}^{a/u} = 2\sin^{-1}\left(\frac{a}{u}\right) = 2\sin^{-1}\left(\frac{a}{\sqrt{1-x^2}}\right).$$

Plug this back into the original integral and get the formula for the surface area:

$$2\int_{-a}^{a}\sin^{-1}\left(\frac{a}{\sqrt{1-x^2}}\right)dx\,.$$

27. (c) Use cylindrical coordinates: $x = 2\cos\theta$, $y = 2\sin\theta$, $z = z$, $0 \le \theta \le 2\pi$, $0 \le z \le x + 3 = 2\cos\theta + 3$. (Note the difference in parametrizations, as compared to exercise 12). $\|\mathbf{T}_\theta \times \mathbf{T}_z\| = \|(-2\sin\theta\,\mathbf{i} + 2\cos\theta\,\mathbf{j})\times\mathbf{k}\| = \|2\cos\theta\,\mathbf{i} + 2\sin\theta\,\mathbf{j}\| = 2$. Then, the surface integral becomes

$$\int_0^{2\pi}\int_0^{2\cos\theta+3} 2z^2\,dz\,d\theta$$

$$= \int_0^{2\pi}\frac{2}{3}(2\cos\theta+3)^3\,d\theta$$

$$= \frac{2}{3}\int_0^{2\pi}(8\cos^3\theta + 36\cos^2\theta + 54\cos\theta + 27)d\theta$$

$$= \frac{2}{3}\int_0^{2\pi}\left[8(1-\sin^2\theta)\cos\theta + 36\left(\frac{1+\cos 2\theta}{2}\right) + 54\cos\theta + 27\right]d\theta$$

$$= \frac{2}{3}\left[8\left(\sin\theta - \frac{\sin^3\theta}{3}\right) + 18\left(\theta + \frac{\sin 2\theta}{2}\right) + 54\sin\theta + 27\,\theta\right]\Bigg|_0^{2\pi}$$

$$= (2/3)[18 \cdot 2\pi + 27 \cdot 2\pi] = 60\pi\,.$$

CHAPTER 8
THE INTEGRAL THEOREMS OF VECTOR ANALYSIS

8.1 Green's Theorem

GOALS

1. Be able to state Green's theorem.

2. Be able to use Green's theorem to compute a line integral or a double integral.

3. Be able to use Green's theorem to find an area.

STUDY HINTS

1. *Notation.* Recall that ∂ , the same symbol used for partial derivatives, means boundary. Thus, ∂D is the boundary of D .

2. *Green's theorem.* Under certain conditions, a line integral is converted into a surface integral:
$$\int_C (P\,dx + Q\,dy) = \int_D \left(\frac{\partial Q}{\partial x} - \frac{\partial P}{\partial y}\right) dx\,dy\,.$$
Often, one side of the equation will be much easier to evaluate than the other side.

3. *Required conditions.* (a) C must be a *closed* curve with a counterclockwise orientation (The region is on your left if you walk around C with the correct orientation), and
(b) P and Q must have continuous first derivatives.
If Green's theorem does not apply directly to a region, the region can usually be subdivided so that the theorem can be applied. See figure 8.1.5 .

4. *Green's theorem and area.* The area of D may be computed by the formula
$$A = \frac{1}{2}\int_{\partial D} (x\,dy - y\,dx)\,.$$
Again, ∂D is oriented counterclockwise. This formula is most useful if the boundary has a simple parametrization; otherwise, double integration is usually simpler.

5. *Vector form.* If $\mathbf{F} = P\mathbf{i} + Q\mathbf{j}$, then $\nabla \times \mathbf{F} = (\partial Q/\partial x - \partial P/\partial y)\,\mathbf{k}$, so another formulation for Green's theorem becomes $\int_{\partial D} \mathbf{F} \cdot d\mathbf{s} = \int_D (\nabla \times \mathbf{F}) \cdot \mathbf{k}\; dA$, where $d\mathbf{s} = dx\mathbf{i} + dy\mathbf{j}$.

6. *Divergence theorem in the plane.* If $\mathbf{n}$ is an outward unit normal to ∂D , then $\int_{\partial D} \mathbf{F} \cdot d\mathbf{s} = \int_D \text{div}\,\mathbf{F}\; dA$. Compare this with Gauss' divergence theorem in space

(section 8.4).

SOLUTIONS TO SELECTED EXERCISES

2. According to Green's theorem, the area of a region D is
$$A = \frac{1}{2}\int_{\partial D} (x\,dy - y\,dx)\,.$$
Since D is a disk centered at (0, 0) with radius R , the boundary ∂D can be parametrized by
$$x = R\cos\theta \quad \text{and} \quad y = R\sin\theta\,,\ 0 \le \theta < 2\pi\,.$$
Then, $dx = -R\sin\theta\,d\theta$, $dy = R\cos\theta\,d\theta$, and so the area is
$$A = \frac{1}{2}\int_0^{2\pi} [(R\cos\theta)(R\cos\theta) - (R\sin\theta)(-R\sin\theta)]d\theta = \frac{1}{2}\int_0^{2\pi} R^2\,d\theta = \pi R^2\,,$$
which is indeed the area that is calculated by using elementary geometry.

3. (b) We use the same parametrization for ∂D as in exercise 2. The left-hand side of the identity in Green's theorem is
$$\int_{\partial D} (P\,dx + Q\,dy) = \int_{\partial D} (x + y)dx + \int_{\partial D} y\,dy$$
$$= \int_0^{2\pi} (R\cos\theta + R\sin\theta)(-R\sin\theta)\,d\theta + \int_0^{2\pi} (R\sin\theta)\,(R\cos\theta)\,d\theta$$
$$= \int_0^{2\pi} (-R^2\cos\theta\sin\theta - R^2\sin^2\theta + R^2\sin\theta\cos\theta)d\theta$$
$$= -R^2\int_0^{2\pi} \sin^2\theta\,d\theta = -R^2\int_0^{2\pi} \frac{1-\cos 2\theta}{2}\,d\theta = -R^2\pi\,.$$
By Green's theorem, the same integral should be equal to
$$\iint_D \left(\frac{\partial Q}{\partial x} - \frac{\partial P}{\partial y}\right)dx\,dy\,.$$
Again, we use polar coordinates. D can be described by $0 \le r \le R$ and $0 \le \theta \le 2\pi$. Calculating $\partial Q/\partial x = 0$ and $\partial P/\partial y = 1$ and remembering that the Jacobian for polar coordinates is r , the right-hand side of Green's theorem becomes
$$\int_0^{2\pi}\int_0^R (0 - 1)r\,dr\,d\theta = \int_0^{2\pi}\int_0^R -r\,dr\,d\theta = -\pi R^2\,.$$
Therefore, Green's theorem is verified for this case.

5. Green's theorem tells us that the area is
$$A = \frac{1}{2}\int_{\partial D} (x\,dy - y\,dx)\,.$$
From the given parametrization of the cycloid, we compute $dx = a(1 - \cos\theta)\,d\theta$ and $dy = a\sin\theta\,d\theta$. Remember that the x axis forms part of the boundary. This part of the boundary can be described by $x = 2\pi a(1 - u)$ and $y = 0$ for $0 \le u \le 1$. Notice that our cycloid is traversed *clockwise.*, that is, from left to right as in the book, then

back along the x axis from right to left. Since $y = dy = 0$, the desired area is

$$\frac{1}{2}\int_0^{2\pi}[a(\theta - \sin\theta)\cdot a\sin\theta\, d\theta - a(1-\cos\theta)\cdot a(1-\cos\theta)\, d\theta] +$$

$$\frac{1}{2}\int_0^{1}(x\cdot 0 - 0\cdot dx)$$

$$= \frac{1}{2}\int_0^{2\pi}(a^2\theta\sin\theta - a^2\sin^2\theta - a^2 + 2a^2\cos\theta - a^2\cos^2\theta)d\theta$$

$$= \frac{1}{2}\int_0^{2\pi}(a^2\theta\sin\theta - 2a^2 + 2a^2\cos\theta)d\theta\,.$$

Using integration by parts, we get

$$-\frac{a^2}{2}\theta\cos\theta\Big|_0^{2\pi} + \frac{a^2}{2}\int_0^{2\pi}\cos\theta\, d\theta + \frac{1}{2}\int_0^{2\pi}[-2a^2 + 2a^2\cos\theta]\, d\theta$$

$$= -a^2\pi + (1/2)(-2a^2\theta + 2a^2\sin\theta)\Big|_0^{2\pi} = -3a^2\pi\,.$$

To get the correct answer, recall the correct direction should be *counterclockwise*. By simply changing the sign, we get the area spanned by one arc of the cycloid, $3a^2\pi$.

8. Let D be the union of type 3 regions, D_i, where each boundary, ∂D_i, is oriented counterclockwise. Suppose P and Q are continuously differentiable on D. We want to show that the boundary of D is the same as the boundary of all of the D_i. Each intersecting boundary lies on the boundary of two regions. When orientations are considered, contributions to the integral cancel out as in figure 8.1.5. Hence

$$\int_{\partial D}(P\,dx + Q\,dy) = \sum_{i=1}^{n}\int_{\partial D_i}(P\,dx + Q\,dy) = \sum_{i=1}^{n}\iint_{D_i}\left(\frac{\partial Q}{\partial x} - \frac{\partial P}{\partial y}\right)dx\,dy$$

$$= \iint_D\left(\frac{\partial Q}{\partial x} - \frac{\partial P}{\partial y}\right)dx\,dy\,.$$

11. (a) Using polar coordinates, we let $x = r\cos\theta$ and $y = r\sin\theta$. Then D is described by $0 \le r \le 1$ and $0 \le \theta \le 2\pi$. We compute that $\operatorname{div}\mathbf{F} = 1 + 1 = 2$. (If necessary, you should review how to compute a divergence in section 3.4.) Since the Jacobian for polar coordinates is r, we get

$$\int_D \operatorname{div}\mathbf{F}\, dA = \int_0^{2\pi}\int_0^{1}(2)r\, dr\, d\theta = 2\pi\,.$$

On the other hand, we know that the outward unit normal to the unit circle is $\mathbf{n} = (\cos\theta, \sin\theta)$, so

$$\int_{\partial D}\mathbf{F}\cdot\mathbf{n}\, ds = \int_0^{2\pi}(\cos\theta, \sin\theta)\cdot(\cos\theta, \sin\theta)\, ds = \int_0^{2\pi}ds = 2\pi\,.$$

(b) We want to compute

$$\int_{\partial D}\mathbf{F}\cdot\mathbf{n}\, ds\,,$$

where $\mathbf{F} = 2xy\mathbf{i} - y^2\mathbf{j}$. By the divergence theorem, it is equal to

$$\iint_D \text{div } \mathbf{F} \, dA = \iint_D (2y - 2y) \, dx \, dy = 0 .$$

15. Since $\cos^2\theta + \sin^2\theta = 1$, let $x/a = \cos\theta$ and $y/b = \sin\theta$, $0 \le \theta \le 2\pi$. Then $x = a\cos\theta$, $dx = -a\sin\theta \, d\theta$, $y = b\sin\theta$, and $dy = b\cos\theta \, d\theta$. By Green's theorem, the area of the ellipse is

$$\begin{aligned} A &= \frac{1}{2}\int_{\partial D} (x\,dy - y\,dx) \\ &= \frac{1}{2}\int_0^{2\pi} [(a\cos\theta)(b\cos\theta) - (b\sin\theta)(-a\sin\theta)] \, d\theta \\ &= \frac{1}{2}\int_0^{2\pi} ab \, d\theta = ab\pi . \end{aligned}$$

16. In polar coordinates, $x = r\cos\theta$ and $y = r\sin\theta$. Since r is a function of θ, we get $dx = (r'\cos\theta - r\sin\theta)\,d\theta$ and $dy = (r'\sin\theta + r\cos\theta)\,d\theta$ (here r' denotes the derivative of r with respect to θ). Substituting into Green's theorem, we get

$$\begin{aligned} A &= \frac{1}{2}\int_C (x\,dy - y\,dx) \\ &= \frac{1}{2}\int_a^b (r\cos\theta)(r'\sin\theta + r\cos\theta)d\theta - (r\sin\theta)(r'\cos\theta - r\sin\theta)d\theta \\ &= \frac{1}{2}\int_a^b r^2(\cos^2\theta + \sin^2\theta)d\theta = \frac{1}{2}\int_a^b r^2 \, d\theta . \end{aligned}$$

19. To form one loop of the rose, we go from $\theta = 0$ to $\theta = \pi/2$. Using the result of exercise 16, the area is

$$\begin{aligned} &\frac{1}{2}\int_0^{\pi/2} (3\sin 2\theta)^2 \, d\theta = \frac{1}{2}\int_0^{\pi/2} (9\sin^2 2\theta)d\theta = \frac{9}{2}\int_0^{\pi/2}\left(\frac{1+\cos 4\theta}{2}\right)d\theta \\ &= \frac{9}{4}\left(\theta + \frac{\sin 4\theta}{4}\right)\Bigg|_0^{\pi/2} = \frac{9\pi}{8} . \end{aligned}$$

22. Use the given definition for $\partial u/\partial n$ and the divergence theorem in the plane for the region $B = B_p$:

$$\int_{\partial B} \frac{\partial u}{\partial n} \, ds = \int_{\partial B} \nabla u \cdot \mathbf{n} \, ds = \int_B \nabla \cdot (\nabla u) \, dA = \int_B \nabla^2 u \, dA .$$

26. (a) Suppose $u(\mathbf{p})$ is a maximum point on D. From exercise 25, $u(\mathbf{p})$ is the *average value* of u on a disk of radius R centered at p. This is possible only if $u(\mathbf{p}) = u(\mathbf{q})$ for all $\mathbf{q}$ in D. Indeed, if $u(\mathbf{q}) < u(\mathbf{p})$ for some $\mathbf{q}$ in D, there must be a $u(\mathbf{r}) > u(\mathbf{p})$ to maintain the average. Therefore, u must be constant on some disk centered at $\mathbf{p}$.

8.2 Stokes' Theorem

GOALS

1. Be able to state and use Stokes' theorem.

2. Be able to use Stokes' theorem to calculate a line integral on a closed curve or a surface integral over a surface with a closed curve as its boundary.

STUDY HINTS

1. *Review.* Surface integrals are used in this section. You should review section 7.6 if you have forgotten how to calculate a surface integral.

2. *Relation to Green's theorem.* Like Green's theorem, Stokes' theorem converts an integral in one dimension to an integral in two dimensions. Stokes' theorem is a generalization of Green's theorem.

3. *Stokes' theorem for graphs.* If $z = f(x, y)$, then
$$\int_S \text{curl } \mathbf{F} \cdot d\mathbf{S} = \int_{\partial S} \mathbf{F} \cdot d\mathbf{s} .$$
This is a formula you should memorize. As with Green's theorem, ∂S is oriented so that the surface is on your left as you walk around ∂S , which must be a closed curve.

4. *Generalized surfaces.* If the surface S is not the graph of a function, then Stokes' theorem is still true if S can be described using a one-to-one parametrization. ∂D gets mapped to ∂S , so ∂D, the boundary for the region in which the parameters reside should have the correct orientation. For an example when the orientation becomes important, see the solution to exercise 5 of section 8.1.

5. *Application.* According to Stokes' theorem, to evaluate $\int_S \text{curl } \mathbf{F} \cdot d\mathbf{S}$, we can change the surface S to any other surface with the same boundary ∂S . In most cases, we will change to a planar surface. Imagine a loop of wire with an elastic sheet S on it. The surface integral of the curl of a vector field over any surface formed by a deformation of the elastic sheet will be equal to the line integral over the wire (assuming the wire itself cannot be deformed).

6. *Circulation.* You should know that curl $\mathbf{V} \cdot \mathbf{n}$ is the circulation of $\mathbf{V}$ per unit area of surface perpendicular to $\mathbf{n}$.

SOLUTIONS TO SELECTED EXERCISES

1. By Stokes' theorem, we only need to evaluate
$$\int_{\partial S} \mathbf{F} \cdot d\mathbf{s},$$
where ∂S is the circle $x^2 + y^2 = 1$, the boundary of our surface. Parametrize the circle

by polar coordinates, or $x = \cos\theta$, $y = \sin\theta$, $0 \le \theta \le 2\pi$. Then the integral becomes

$$\int_{\partial S} \mathbf{F} \cdot d\mathbf{s} = \int_0^{2\pi} \sin\theta\,(-\sin\theta) - \cos\theta\,(\cos\theta)\, d\theta$$

$$= -\int_0^{2\pi} d\theta = -2\pi\,.$$

4. We want to show that the derivative of the magnetic flux with respect to time is 0. We begin with

$$\frac{\partial}{\partial t}\int_S \mathbf{H} \cdot d\mathbf{S} = \int_S \frac{\partial \mathbf{H}}{\partial t} \cdot d\mathbf{S} = -\int_S -\frac{\partial \mathbf{H}}{\partial t} \cdot d\mathbf{S}\,.$$

The first step is justified because S is not a function of time. By Faraday's law, we then get

$$-\int_S (\nabla \times \mathbf{E}) \cdot d\mathbf{S}\,.$$

Then by Stokes' theorem and the fact that $\mathbf{E} \cdot d\mathbf{s} = 0$, since E is perpendicular to the boundary of S, we get

$$-\int_{\partial S} \mathbf{E} \cdot d\mathbf{s} = 0\,.$$

5. The boundary of the surface is a closed curve, so we may take advantage of Stokes' theorem:

$$\int_S (\nabla \times \mathbf{F}) \cdot d\mathbf{S} = \int_{\partial S} \mathbf{F} \cdot d\mathbf{s}\,.$$

The boundary is the circle $x^2 + y^2 = 1$, $z = 0$. The right-hand side becomes

$$\int_{\partial S} \mathbf{F} \cdot d\mathbf{s} = \int_{\partial S} (x, y) \cdot (dx, dy)\,.$$

Now use polar coordinates: let $x = \cos\theta$, $dx = -\sin\theta\, d\theta$; $y = \sin\theta$, $dy = \cos\theta\, d\theta$; $0 \le \theta \le 2\pi$. Substitution into the last integral gives us

$$\int_0^{2\pi} (-\cos\theta \sin\theta + \sin\theta \cos\theta) d\theta = 0\,.$$

9. Use Stokes' theorem. The boundary of S is the circle $y^2 + z^2 = 1$, $x = 0$. On the boundary, $\mathbf{F}$ becomes $-y^3\mathbf{j}$. Use polar coordinates: $y = \sin\theta$, $z = \cos\theta$, $0 \le \theta \le 2\pi$. By Stokes' theorem, we get

$$\iint_S (\nabla \times \mathbf{F}) \cdot d\mathbf{S} = \int_{\partial S} \mathbf{F} \cdot d\mathbf{s} = \int_0^{2\pi} -\sin^3\theta \cos\theta\, d\theta = \frac{-1}{4}\sin^4\theta \Big|_0^{2\pi} = 0\,.$$

12. The flow rate is $\int_S \operatorname{curl} \mathbf{\Phi} \cdot d\mathbf{S}$ and by Stokes' theorem, this is $\int_{\partial S} \mathbf{\Phi} \cdot d\mathbf{s}$. We have the boundary on the xy plane where $z = 0$. Parametrize the boundary by

$$x = (R/4)\cos\theta\,,\ y = (R/4)\sin\theta\,.$$

Then the flux is

$$\int_{\partial S} \mathbf{\Phi} \cdot d\mathbf{s} = \int_0^{2\pi} \frac{R^2}{16} (\sin^2\theta + \cos^2\theta)\, d\theta = \frac{\pi R^2}{8} .$$

14. By Stokes' theorem,

$$\int_S (\nabla \times \mathbf{F}) \cdot d\mathbf{S} = \int_{\partial S} \mathbf{F} \cdot d\mathbf{s} .$$

Since $\mathbf{F}$ is perpendicular to the tangent to the boundary of S, $\mathbf{F} \cdot d\mathbf{s} = 0$. Hence $\int_S (\nabla \times \mathbf{F}) \cdot d\mathbf{s} = 0$. If $\mathbf{F}$ is an electric field, this means the rate of change of magnetic flux is zero by Faraday's law. See Example 4.

21. First, we will calculate $\int_S (\nabla \times \mathbf{F}) \cdot d\mathbf{S}$. We compute $\nabla \times \mathbf{F} = (1, 1, 1)$. Also, $\mathbf{\Phi}_r = (\cos\theta, \sin\theta, 0)$ and $\mathbf{\Phi}_\theta = (-r\sin\theta, r\cos\theta, 1)$, so $\mathbf{\Phi}_r \times \mathbf{\Phi}_\theta =$ $(\sin\theta, -\cos\theta, r)$. Therefore, we get

$$\int_S (1, 1, 1) \cdot (\sin\theta, -\cos\theta, r)\, dr\, d\theta$$

$$= \int_0^1 \int_0^{\pi/2} (\sin\theta - \cos\theta + r)\, d\theta\, dr = \int_0^1 \left(\frac{\pi}{2} r\right) dr = \frac{\pi}{4} .$$

On the other hand, the boundary, ∂S, is composed of four parts. First, when $r = 1$, we have $\mathbf{\Phi}(1, \theta) = (\cos\theta, \sin\theta, \theta)$, so $\mathbf{F} = (\theta, \cos\theta, \sin\theta)$ and $d\mathbf{s} = d\mathbf{\Phi}(1, \theta) =$ $(-\sin\theta, \cos\theta, 1)\, d\theta$. Therefore,

$$\int_{\partial S_1} \mathbf{F} \cdot d\mathbf{s} = \int_0^{\pi/2} (\theta, \cos\theta, \sin\theta) \cdot (-\sin\theta, \cos\theta, 1)\, d\theta .$$

Using integration by parts to integrate $-\theta\sin\theta$ and the half-angle formula to integrate $\cos^2\theta$, we get

$$\left[(\theta\cos\theta - \sin\theta) + \left(\frac{\theta}{2} + \frac{\sin 2\theta}{4}\right) - \cos\theta \right] \Bigg|_0^{\pi/2} = \frac{\pi}{4} .$$

When $\theta = \pi/2$, orientation is maintained by letting r go from 1 to 0. Thus, we have

$$\int_{\partial S_2} \mathbf{F} \cdot d\mathbf{s} = \int_1^0 \left(\frac{\pi}{2}, 0, r\right) \cdot (0, 1, 0)\, dr = 0 .$$

When $r = 0$, θ goes from $\pi/2$ to 0, so we get

$$\int_{\partial S_3} \mathbf{F} \cdot d\mathbf{s} = \int_{\pi/2}^0 (\theta, 0, 0) \cdot (0, 0, 1)\, d\theta = 0 .$$

Similarly, when $\theta = 0$, we get

$$\int_{\partial S_4} \mathbf{F} \cdot d\mathbf{s} = \int_0^1 (0, r, 0) \cdot (1, 0, 0)\, dr = 0 .$$

Adding all the parts together, we get $\int_{\partial S} \mathbf{F} \cdot d\mathbf{s} = \pi/4 + 0 + 0 + 0 = \pi/4$, so theorem 6 is verified.

25. For a direct computation, parametrize the surface as follows: Let $x = r\cos\theta$ and $y = r\sin\theta$, so $z = (1/2)(x^2 + y^2) = r^2/2$. Also, we want $0 \le z \le 2$, so $0 \le r^2/2 \le 2$, or $0 \le r \le 2$. In addition, we have $0 \le \theta \le 2\pi$. We calculate $\mathbf{T}_\theta =$ $(-r\sin\theta, r\cos\theta, 0)$ and $\mathbf{T}_r = (\cos\theta, \sin\theta, r)$, so the outward normal is $\mathbf{T}_\theta \times \mathbf{T}_r =$ $(-r^2\cos\theta, -r^2\sin\theta, -r)$. Also, we calculate

$$\nabla \times \mathbf{F} = \begin{vmatrix} \mathbf{i} & \mathbf{j} & \mathbf{k} \\ \partial/\partial x & \partial/\partial y & \partial/\partial z \\ 3y & -xz & yz^2 \end{vmatrix} = (z^2 - x, 0, -z - 3) = \left(\frac{1}{4}r^4 - r\cos\theta, 0, \frac{-1}{2}r^2 - 3\right).$$

Finally, $\int_S (\nabla \times \mathbf{F}) \cdot d\mathbf{S}$ becomes

$$\int_0^2 \int_0^{2\pi} (\nabla \times \mathbf{F}) \cdot (\mathbf{T}_\theta \times \mathbf{T}_r)\, d\theta\, dr$$

$$= \int_0^2 \int_0^{2\pi} \left[-r^2\cos\theta\left(\frac{1}{4}r^4 - r\cos\theta\right) + r\left(\frac{1}{2}r^2 + 3\right)\right] d\theta\, dr$$

$$= \int_0^2 \int_0^{2\pi} \left[\frac{-r^6}{4}\cos\theta + r^3\cos^2\theta + \frac{1}{2}r^3 + 3r\right] d\theta\, dr$$

$$= -\int_0^2 \left[\left(\frac{r^6}{4}\sin\theta + r^3\left(\frac{\theta}{2} + \frac{\sin 2\theta}{4}\right) - \left(\frac{1}{2}r^3 + 3r\right)\theta\right)\Big|_{\theta=0}^{2\pi}\right] dr$$

$$= \int_0^2 (2\pi r^3 + 6\pi r)\, dr = \left(\frac{\pi r^4}{2} + 3\pi r^2\right)\Big|_0^2 = 20\pi.$$

On the other hand, by Stokes' theorem, $\int_S (\nabla \times \mathbf{F}) \cdot d\mathbf{S} = \int_{\partial S} \mathbf{F} \cdot d\mathbf{s}$. The boundary is ∂S, which is the circle of radius 2 at $z = 2$. It can be parametrized by $(2\cos t, -2\sin t, 2)$ for $0 \le t \le 2\pi$. We use this orientation because the surface lies below the boundary, so we should traverse it in a *clockwise* orientation. We compute $d\mathbf{s} = (-2\sin t, -2\cos t, 0)dt$, so

$$\int_{\partial S} \mathbf{F} \cdot d\mathbf{s} = \int_0^{2\pi} (-6\sin t, -4\cos t, -8\sin t) \cdot (-2\sin t, -2\cos t, 0)\, dt$$

$$= \int_0^{2\pi} (12\sin^2 t + 8\cos^2 t)\, dt$$

$$= \left[12\left(\frac{t}{2} - \frac{\sin 2t}{4}\right) + 8\left(\frac{t}{2} + \frac{\sin 2t}{4}\right)\right]\Big|_0^{2\pi} = 20\pi.$$

8.3 Conservative Fields

GOALS

1. Understand that the line integral of a gradient field is path independent.
2. Be able to determine whether a vector field is conservative.
3. Given a conservative vector field, be able to find a scalar function whose gradient is equal to that vector field.

STUDY HINTS

1. *Theorem 7.* This is a very important theorem. To summarize, it states that if $\mathbf{F}$ is a gradient, then $\nabla \times \mathbf{F} = \mathbf{0}$, the line integral depends only on the endpoints, and all line integrals around closed curves are 0. The converse is also true. If the theorem holds, then we say that the line integrals are "independent of path." Note that if a single line integral is 0, then $\mathbf{F}$ may not necessarily be a gradient.

2. *Example 1.* In method 2, part (a), notice how after integrating in x, we add a "constant" $h_1(y, z)$. It is a "constant" because it only involves the other variables. Study example 1 carefully. You should know how to use at least one of the two methods.

3. *Gradients in* $\mathbf{R}^2$. The corollary preceding example 3 tells you that if the integrand in Green's theorem, $\partial Q/\partial x - \partial P/\partial y$, is 0, then $\mathbf{F} = P\mathbf{i} + Q\mathbf{j}$ is a gradient. Be careful!! $\mathbf{F}$ must be C^1 on *all* of $\mathbf{R}^2$, unlike theorem 7, which allows some exceptional points. Exercise 12 elaborates this point. The integral of $\mathbf{F}$ over the path on the left (a closed path) is 0 because the path does not include the origin, while the integral of $\mathbf{F}$ over the path on the right is not.

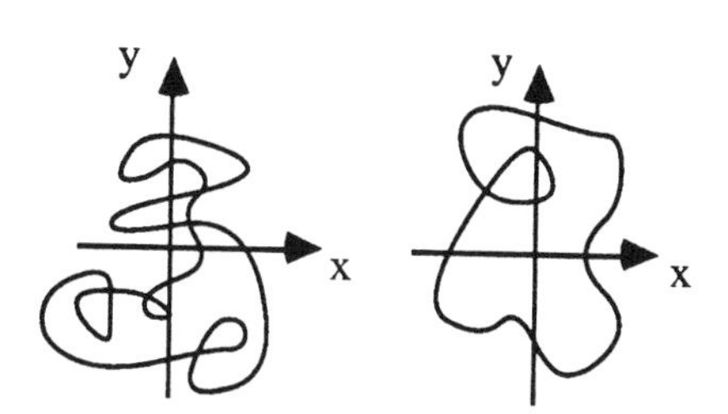

4. *Is* $\mathbf{F}$ *a curl?* $\mathbf{F}$ is a curl of some vector field if $\operatorname{div} \mathbf{F} = 0$. Exercise 16 explains the procedure for finding a $\mathbf{G}$ such that $\mathbf{F} = \operatorname{curl} \mathbf{G}$.

SOLUTIONS TO SELECTED EXERCISES

2. (a) Since $y = 2x^2$, we have $dy = 4x\,dx$, and so

$$\int_\sigma \mathbf{F} \cdot d\mathbf{s} = \int_0^1 (x \cdot 2x^2)\,dx + (2x^2)^2 \cdot 4x\,dx = \int_0^1 (2x^3 + 16x^5)\,dx = \frac{1}{2} + \frac{8}{3} = \frac{19}{6}.$$

(b) The answer is yes, since

$$\nabla \times \mathbf{F} = \begin{vmatrix} \mathbf{i} & \mathbf{j} & \mathbf{k} \\ \partial/\partial x & \partial/\partial y & \partial/\partial z \\ xy & y^2 & 0 \end{vmatrix} = (0, 0, -x) \neq \mathbf{0} .$$

Alternatively, one can pick a different path and show that the line integral is not equal to $19/6$. This would mean that the line integral of $\mathbf{F}$ is path dependent.

3. If $\mathbf{F} = \nabla f$, then the x component of $\mathbf{F}$ must be $\partial f/\partial x$, i.e.,

$$\partial f/\partial x = 2xyz + \sin x . \qquad (1)$$

Similarly, the y component of $\mathbf{F}$ must be $\partial f/\partial y$ and the z component of $\mathbf{F}$ must be $\partial f/\partial z$, i.e.,

$$\partial f/\partial y = x^2 z \qquad (2)$$
$$\partial f/\partial z = x^2 y . \qquad (3)$$

Integrating (1) with respect to x, we get $f = \int (2xyz + \sin x)\, dx = x^2yz - \cos x + h(y, z)$, where h is a function of y and z only. When we integrate with respect to x we should restore all the terms with x. The terms without x are treated as a constant when differentiated, and integration with respect to x cannot restore them. Similarly, integrate (2) with respect to y to get $f = \int x^2 z\, dy = x^2zy + g(x, z)$, where $g(x, z)$ is a function of x and z only. Integrating (3) with respect to z gives us $f = \int x^2 y\, dz = x^2yz + k(x, y)$, where $k(x, y)$ is a function of x and y only. Compare the three results: $f(x, y, z) = x^2yz - \cos x + h(y, z) = x^2yz + g(x, z) = x^2yz + k(x, y)$. We conclude (by inspection) that $g(x, z) = k(x, y) = -\cos x + C$ and $h(y, z) = C$, where C is a constant. Thus,

$$f(x, y, z) = x^2yz - \cos x + C .$$

6. We use the chain rule to compute $(\partial/\partial x)(1/\sqrt{x^2 + y^2 + z^2}) = -x/(x^2 + y^2 + z^2)^{3/2}$. Then, by symmetry, we get

$$\nabla\left(\frac{1}{r}\right) = \nabla\left(\frac{1}{\sqrt{x^2 + y^2 + z^2}}\right) = \frac{-(x\mathbf{i} + y\mathbf{j} + z\mathbf{k})}{(x^2 + y^2 + z^2)^{3/2}} = \frac{-\mathbf{r}}{\|\mathbf{r}\|^3} .$$

Since $\mathbf{F}$ is the gradient of a function f, $\int_{\boldsymbol{\sigma}} \mathbf{F} \cdot d\mathbf{s} = f(\boldsymbol{\sigma}(a)) - f(\boldsymbol{\sigma}(b))$ for all paths $\boldsymbol{\sigma}(s)$ beginning at a and ending at b. This is only true if $\boldsymbol{\sigma}$ does not pass through the origin.

9. The reader should verify that $\nabla \times \mathbf{F} = \mathbf{0}$. Hence $\mathbf{F}$ is a gradient of some function $f(x, y, z)$. Integrate the $\mathbf{i}$ component with respect to x, the $\mathbf{j}$ component with respect to y, and the $\mathbf{k}$ component with respect to z. Comparing the results, we see that $f(x, y, z) = e^x \sin y + z^3/3$. Also, we compute $\boldsymbol{\sigma}(0) = (0, 0, 1)$ and $\boldsymbol{\sigma}(1) = (1, 1, e)$. Since $\mathbf{F}$ is a gradient, we have $\int_{\boldsymbol{\sigma}} \mathbf{F} \cdot d\mathbf{s} = f(\boldsymbol{\sigma}(1)) - f(\boldsymbol{\sigma}(0)) = e \sin 1 + e^3/3 - 1/3$.

13. (b) $\mathbf{F}$ is *not* the gradient of a scalar function f. If such f were to exist, then $\partial f/\partial x = xy$ and $\partial f/\partial y = xy$. By the equality of mixed partials, we would expect that $\partial^2 f/\partial x\partial y = \partial^2 f/\partial y\partial x$. But in this case, $(\partial/\partial y)(\partial f/\partial x) = x$ and $(\partial/\partial x)(\partial f/\partial y) = y$.

15. (b) To show that $\mathbf{\Phi}$ is conservative, we can show that it is a gradient. (We can also show that any one of the four conditions in theorem 7 is met). We will use the same method as in exercise 13. The partial derivatives are:

$$\frac{\partial}{\partial y}\left(\frac{2x}{y^2+1}\right) = \frac{(2x)(-2y)}{(y^2+1)^2} = \frac{-4xy}{(y^2+1)^2};$$

$$\frac{\partial}{\partial x}\left(\frac{-2y(x^2+1)}{(y^2+1)^2}\right) = \frac{-4xy}{(y^2+1)^2}.$$

Since these partials are equal, $\mathbf{\Phi}$ is conservative. This makes the evaluation of the path integral easy. You should verify that if $f(x, y) = (x^2 + 1)/(y^2 + 1)$, then $\mathbf{\Phi} = \nabla f$. Then substitute $x = t^3 - 1$ and $y = t^6 - t$ to get $f(t) = [(t^3 - 1)^2 + 1]/[(t^6 - t)^2 + 1]$, and so

$$\int_C \mathbf{\Phi} \cdot d\mathbf{s} = \left[\frac{(t^3-1)^2+1}{(t^3-t)^2+1}\right]\Bigg|_0^1 = -1.$$

18. First, we compute $\nabla \cdot \mathbf{F} = (\partial/\partial x)xz + (\partial/\partial y)(-yz) + (\partial/\partial z)y = z - z + 0 = 0$. Thus, there exists a $\mathbf{G}$ such that $\mathbf{F} = \nabla \times \mathbf{G}$. To find $\mathbf{G}$, use the result of exercise 16:

$$G_1 = \int_0^z -yt\, dt - \int_0^y t\, dt = \frac{-yz^2}{2} - \frac{y^2}{2},$$

$$G_2 = -\int_0^z xt\, dt = \frac{-xz^2}{2}, \text{ and } G_3 = 0.$$

Hence $\mathbf{G} = (-1/2)[(yz^2 + y^2)\mathbf{i} + xz^2\mathbf{j}]$. It is a good idea to compute curl $\mathbf{G}$ to verify your answer:

$$\nabla \times \mathbf{G} = \frac{-1}{2}\begin{vmatrix} \mathbf{i} & \mathbf{j} & \mathbf{k} \\ \partial/\partial x & \partial/\partial y & \partial/\partial z \\ yz^2+y^2 & xz^2 & 0 \end{vmatrix} = \frac{-1}{2}(-2xz\mathbf{i} + 2yz\mathbf{j} + (z^2 - z^2 - 2y)\mathbf{k})$$
$$= xz\mathbf{i} - yz\mathbf{j} + y\mathbf{k} = \mathbf{F}.$$

Note that $\mathbf{G}$ is not unique; for example, arbitrary constants may be added to each component of $\mathbf{G}$, and $\nabla \times \mathbf{G}$ would still be equal to $\mathbf{F}$.

23. (a) Recall from chapter 3 that $\mathbf{F}$ is not irrotational means that curl $\mathbf{F} \neq \mathbf{0}$. Thus, we compute curl $\mathbf{F}$:

$$\nabla \times \mathbf{F} = \begin{vmatrix} \mathbf{i} & \mathbf{j} & \mathbf{k} \\ \partial/\partial x & \partial/\partial y & \partial/\partial z \\ -y & x & 0 \end{vmatrix} = 2\mathbf{k} \neq \mathbf{0}.$$

(b) Let $\boldsymbol{\sigma}(t) = (x(t), y(t), z(t))$ be the trajectory of the cork. By the definition of flow lines, $\boldsymbol{\sigma}'(t) = \mathbf{F}(\boldsymbol{\sigma}(t))$, and so $\boldsymbol{\sigma}'(t) = (-y(t), x(t), 0)$. Equivalently, we have the following system of differential equations:

$$x'(t) = -y(t) \qquad (1)$$
$$y'(t) = x(t) \qquad (2)$$
$$z'(t) = 0 \qquad (3)$$

Equation (3) can be solved easily. Its solution is $z(t) = \text{constant}$. Taking one more derivative of (1) with respect to t yields $x''(t) = -y'(t)$, but (2) implies that $x''(t) = -x(t)$, or $x''(t) + x(t) = 0$. This equation represents a harmonic oscillator. The famous solution to the differential equation $x''(t) + x(t) = 0$ is

$$x(t) = A \sin t + B \cos t,$$

where A and B are constants. The reader may verify this is indeed the solution by substitution, if the reader is unfamiliar with the techniques of solving ordinary differential equations. Similarly,

$$y(t) = C \sin t + D \cos t,$$

where C and D are constants. Since $x''(t) = -y'(t)$, we differentiate x and y and compare terms to find that $C = B$ and $D = -A$. Therefore, $y(t) = B \sin t - A \cos t$. Squaring x and y and adding them gives us $x^2 + y^2 = A^2 \sin^2 t + 2AB \sin t \cos t + B^2 \cos^2 t + B^2 \sin^2 t - 2AB \sin t \cos t + A^2 \cos^2 t = A^2 + B^2$. Since $A^2 + B^2$ is a constant, we recognize the equation $x^2 + y^2 = A^2 + B^2$ as the equation of a circle of radius $\sqrt{A^2 + B^2}$ centered at $(0, 0)$. Thus, the cork has a circular trajectory about the z axis in a plane parallel to the xy plane.

(c) As y increases, x decreases, since $x'(t) = -y$. We also know that the cork is going in a circle. So the cork is revolving counterclockwise.

24. (c) The property of being rotational is a local property, that is, the field is rotational at point. In exercise 23, the cork whirls while going around in a circle, but in exercise 24, the cork does not. Single trajectories have little to do with the rotationalness of the fluid.

8.4 Gauss' Theorem

GOALS

1. Be able to state and use Gauss' theorem.

2. Be able to use Gauss' theorem to compute a double integral over a closed surface or a triple integral over a volume enclosed by a surface.

STUDY HINTS

1. *Definition.* A *closed surface* is a surface which must be punctured in order to get into the region it encloses. The enclosed region is denoted Ω and the closed surface is denoted $\partial\Omega$.

2. *Gauss' divergence theorem.* If $\partial\Omega$ is a closed surface, then

$$\int_{\Omega} (\text{div } \mathbf{F})\, dV = \int_{\partial\Omega} (\mathbf{F} \cdot \mathbf{n})\, dS .$$

Thus, a triple integral is reduced to a double integral, or vice versa. Compare this to the divergence theorem in the plane (section 8.1).

3. *Physical interpretation.* div $\mathbf{F}(P)$ is the outward flow at the point P per unit volume. If div $\mathbf{F}(P) > 0$, material flows out, so P is called a source. P is a sink if div $\mathbf{F}(P) < 0$. If div $\mathbf{F}(P) = 0$, the vector field is divergence free, that is, what goes in must come out.

SOLUTIONS TO SELECTED EXERCISES

4. (a) For the faces parallel to the yz plane, the outward pointing unit normal vectors are $\mathbf{i}$ and $-\mathbf{i}$, respectively. For those two faces, we have

$$\int_{\partial\Omega} \mathbf{F} \cdot d\mathbf{S} = \int_0^1\int_0^1 (\mathbf{i}+\mathbf{j}+\mathbf{k})\cdot \mathbf{i}\; dy\, dz + \int_0^1\int_0^1 (\mathbf{i}+\mathbf{j}+\mathbf{k})\cdot(-\mathbf{i})\; dy\, dz$$

$$= \int_0^1\int_0^1 (\mathbf{i}+\mathbf{j}+\mathbf{k})\cdot(\mathbf{i}-\mathbf{i})\; dy\, dz = 0 .$$

The outward pointing unit normal vectors $\mathbf{n}$ for any two parallel faces are the exact opposite, so the integrals over any two parallel faces of the cube cancel. Therefore, the integral is 0. Next, we calculate that div $\mathbf{F} = 0$, so by the divergence theorem, the desired integral is

$$\int_{\Omega} \text{div } \mathbf{F}\; dV = 0 .$$

6. (b) First, we see that div $\mathbf{F} = 1 + 1 + 1 = 3$, so by the divergence theorem,

$$\int_{\partial\Omega} \mathbf{F} \cdot d\mathbf{S} = \int_{\Omega} \text{div } \mathbf{F}\; dV = 3\int_{\Omega} dV ,$$

which is 3 times the volume of Ω. The problem now is to find the volume of the region of interest. Use "cylindrical" coordinates. We will let θ range from $-\pi/2$ to $\pi/2$ since $x \geq 0$. In addition, we have $0 \leq r \leq 1$ and $x^2 + y^2 = r^2 \leq z \leq 1$. Therefore,

$$\int_{\Omega} dV = \int_{-\pi/2}^{\pi/2}\int_0^1\int_{r^2}^1 r\; dz\, dr\, d\theta = \pi\int_0^1 r(1-r^2)\; dr = \pi\left(\frac{1}{2}-\frac{1}{4}\right) = \frac{\pi}{4} .$$

So

$$\int_{\partial\Omega} \mathbf{F} \cdot ds = 3\left(\frac{\pi}{4}\right) = \frac{3\pi}{4} .$$

10. Use the divergence theorem. We have div $\mathbf{F} = (x^2 + y^2)^2$, so

$$\iint_{\partial S} \mathbf{F} \cdot d\mathbf{S} = \iiint_S (x^2+y^2)^2\; dV .$$

Use cylindrical coordinates. The cylinder S can be described by $0 \leq r \leq 1$, $0 \leq \theta \leq 2\pi$, and $0 \leq z \leq 1$. Since the Jacobian is r, we get

$$\iint_{\partial S} \mathbf{F} \cdot d\mathbf{S} = \int_0^{2\pi} \int_0^1 \int_0^1 r \cdot r^4 \, dr \, dz \, d\theta = 2\pi \int_0^1 r^6 \, dr = \frac{\pi}{3} .$$

16. From section 8.3, we know that curl $\mathbf{F} = \mathbf{0}$ implies $\mathbf{F} = \nabla f$, so taking the divergence of both sides gives us div $\mathbf{F} = \nabla \cdot (\nabla f) = \nabla^2 f$. But we are given that div $\mathbf{F} = 0$. Hence $\nabla^2 f = 0$.

18. By the divergence theorem, $\int_\Omega \text{div}\, \mathbf{F}\, dV = \int_{\partial\Omega} \mathbf{F} \cdot d\mathbf{S}$. Given that $\mathbf{F}$ is tangent to the surface, we know that $\mathbf{F}$ is perpendicular to the unit normal of the surface S, and so $\mathbf{F} \cdot d\mathbf{S} = 0$. Hence $\int_\Omega \text{div}\, \mathbf{F}\, dV = 0$.

8.5 Applications to Physics and Differential Equations

GOALS

1. Be able to use vector analysis techniques in applications such as electromagnetism and Green's function.

STUDY HINTS

1. *Notation.* There's a lot of new notation, so you may find it useful to make a list of new symbols and their definitions.

2. *Old concepts.* No new mathematical material is presented in this section. Only the terminology has changed. You should be able to justify each step with the knowledge that you have acquired in the previous chapters.

3. *Dirac delta function.* The Dirac delta "function" is a very important tool in physical sciences. Its most important properties are that it can "pick out" the value of a function at a particular point and that it "peaks" out at a point.

4. *Green's function.* Green's function is an important tool in solving inhomogeneous differential equations with boundary conditions. It gives the solution if all the "stuff" (charge, mass, etc.) is concentrated at one point; tthrough integration, it gives the solution that solves the equation for a particular distribution of the stuff.

SOLUTIONS TO SELECTED EXERCISES

4. Given $\nabla^2 \varphi = 0$ and $\mathbf{V} = \nabla \varphi$, we want to show that $\rho(\partial \mathbf{V}/\partial t + \mathbf{V} \cdot \nabla \mathbf{V}) = -\nabla p$ if $\partial \mathbf{V}/\partial t = 0$. Using exercise 3, $\mathbf{V} \cdot \nabla \mathbf{V} = (1/2)\nabla(\| \mathbf{V} \|^2) + (\nabla \times \mathbf{V}) \times \mathbf{V}$. Since $\mathbf{V}$ is

a gradient, $\nabla \times \mathbf{V} = 0$. It suffices to show that $\rho[(1/2)\nabla(\| \mathbf{V} \|^2)] = -\nabla p$ .Choose $p = -(\rho/2)\| \mathbf{V} \|^2$.

5. Start with Ampere's law: $\nabla \times \mathbf{H} - \partial \mathbf{E}/\partial t = \mathbf{J}$. Take the divergence of both sides: $\operatorname{div} \mathbf{J} = \nabla \cdot (\nabla \times \mathbf{H}) - \nabla \cdot (\partial/\partial t)\mathbf{E} = 0 - (\partial/\partial t)(\nabla \cdot \mathbf{E})$ (the change of order of the two derivatives is justified since ∇ takes the derivatives of spatial variables only) . Now, by Gauss' law, $\nabla \cdot \mathbf{E} = \rho$. Combine the two results, we get $\operatorname{div} \mathbf{J} = -\partial\rho/\partial t$, or $\operatorname{div} \mathbf{J} + \partial\rho/\partial t = 0$, which is the equation of continuity.

6. (a) According to exercise 12 on page 552, we need to show that

   (1) $\tilde{G}(\mathbf{x}, \mathbf{y}) = \tilde{G}(\mathbf{y}, \mathbf{x})$ and

   (2) $\nabla^2 G = \delta(\mathbf{x}, \mathbf{y})$.

   Define $\mathbf{r} = \mathbf{x} - \mathbf{y}$ and $\mathbf{r}' = \mathbf{R}(\mathbf{x}) - \mathbf{y}$. To show (1), $\tilde{G}(\mathbf{y}, \mathbf{x}) = G(\mathbf{y}, \mathbf{x}) - G(\mathbf{y}, R(\mathbf{x})) = -1/4\pi \| \mathbf{y} - \mathbf{x} \| + 1/4\pi \| \mathbf{y} - R(\mathbf{x}) \|$. It is obvious that $\| \mathbf{r} \| = \| -\mathbf{r} \|$, so $\tilde{G}(\mathbf{y}, \mathbf{x}) = \tilde{G}(\mathbf{x}, \mathbf{y})$. To show (2) , we know $\nabla G(\mathbf{x}, \mathbf{y}) = \mathbf{r}/4\pi r^3$. Thus $\nabla G(R(\mathbf{x}), \mathbf{y}) = \mathbf{r}'/4\pi(r')^3$. Then $\nabla^2\tilde{G} = \delta(\mathbf{x}, \mathbf{y}) - \delta(R(\mathbf{x}), \mathbf{y})$, and

   $$\int_{\mathbf{R}^3} \nabla^2\tilde{G}\, d\mathbf{y} = \int_{\mathbf{R}^3} [\delta(\mathbf{x}, \mathbf{y}) - \delta(R(\mathbf{x}), \mathbf{y})]\, d\mathbf{y} .$$

   If $\mathbf{x} = \mathbf{y}$, then $R(\mathbf{x}) \neq \mathbf{y}$, and the above integral becomes

   $$\int_{\mathbf{R}^3} \nabla^2\tilde{G}\, d\mathbf{y} = \int_{\mathbf{R}^3} \delta(\mathbf{x}, \mathbf{y})\, d\mathbf{y} = 1 ;$$

   If $R(\mathbf{x}) = \mathbf{y}$, then $\mathbf{x} \neq \mathbf{y}$. Make a change of variable: Let $u = x$, $v = y$, $w = -z$, so $d\mathbf{y} = dx\, dy\, dz = -du\, dv\, dw$. Now the integral becomes

   $$\int_{\mathbf{R}^3} \nabla^2\tilde{G}\, d\mathbf{y} = -\int_{\mathbf{R}^3} \delta(R(\mathbf{x}), \mathbf{y})\, d\mathbf{y}$$

   $$= -\int_{\mathbf{R}^3} \delta(R(\mathbf{x}), \mathbf{u})\, (-d\mathbf{u}) = \int_{\mathbf{R}^3} \delta(R(\mathbf{x}), \mathbf{u})\, d\mathbf{u} = 1 ,$$

   where $\mathbf{u} = (u, v, w)$.

   (b) Simply "stick" our Green's function into an integral:

   $$u(\mathbf{x}) = \int_{\mathbf{R}^3} \tilde{G}(\mathbf{x}, \mathbf{y})\, \rho(\mathbf{y})\, d\mathbf{y} = \int_{\mathbf{R}^3} G(\mathbf{x}, \mathbf{y})\, \rho(\mathbf{y}) - G(R(\mathbf{x}), \mathbf{y})\, \rho(\mathbf{y})\, d\mathbf{y} .$$

8. (a) Use the chain rule: $u_t + u_{xx} = u_t \dot{t} + u_x \dot{x} = \dot{u} = 0$.

   (b) Use $dt/ds = 1$ and $dx/ds = u$. The slope is equal to

   $$\frac{t(s)}{x(s)} = \frac{dt}{dx} = \frac{1}{u} .$$

   From part (a), u is a constant, therefore $1/u$ is also a constant. Hence the characteristic curves are straight lines.

   (c) Two characteristics through $(x_1, 0)$ and $(x_2, 0)$ have equations $t = [1/u_0(x_1)](x - x_1)$ and $t = [1/u_0(x_2)](x - x_2)$, respectively. The intersection is

$$[1/u_0(x_1)](x - x_1) = [1/u_0(x_2)](x - x_2) .$$

Simplify and get

$$x(1/u_0(x_1) - 1/u_0(x_2)) = x_1/u_0(x_1) - x_2/u_0(x_2) .$$

Solve for x :

$$x = \frac{\dfrac{x_1}{u_0(x_1)} - \dfrac{x_2}{u_0(x_2)}}{\dfrac{1}{u_0(x_1)} - \dfrac{1}{u_0(x_2)}} = \frac{x_1u_0(x_2) - x_2u_0(x_1)}{u_0(x_2) - u_0(x_1)} > 0 .$$

(d) Plug in:

$$\bar{t} = \frac{1}{u_0(x_1)}\left(\frac{x_1u_0(x_2) - x_2u_0(x_1)}{u_0(x_2) - u_0(x_1)} - x_1\right) = -\frac{x_2 + x_1}{u_0(x_2) - u_0(x_1)} .$$

16. (a) By definition, the Poynting vector is

$$\int_S \mathbf{P} \cdot d\mathbf{S} = \int_S (\mathbf{E} \times \mathbf{H}) \cdot d\mathbf{S}.$$

By Gauss' theorem, we may rewrite the right hand side as

$$\int_V \nabla \cdot (\mathbf{E} \times \mathbf{H})\, dV .$$

The integrand can be rewritten according to the property of triple product (see chapter 1):

$$\int_V \nabla \cdot (\mathbf{E} \times \mathbf{H})\, dV = -\int_V \mathbf{E} \cdot (\nabla \times \mathbf{H})\, dV .$$

Finally, Ampere's law gives, for a static electric field,

$$\nabla \times \mathbf{H} = \mathbf{E} \cdot \mathbf{J} .$$

Combine everything together, we get

$$\int_S \mathbf{P} \cdot d\mathbf{S} = -\int_V \mathbf{E} \cdot \mathbf{J}\, dV .$$

8.6 Differential Forms

GOALS

1. Be able to add and wedge multiply k-forms.

2. Be able to integrate and differentiate k-forms.

3. Be able to state Green's, Stokes', and Gauss' theorems in terms of k-forms.

STUDY HINTS

1. *Notation.* In this textbook, ω is used for 1-forms, η is used for 2-forms, and ν is used for 3-forms.

2. *Definitions.* (a) 0-forms are scalar functions.
(b) 1-forms contain dx, dy, or dz.
(c) 2-forms contain $dx\,dy$, $dy\,dz$, or $dz\,dx$ in the order specified.
(d) 3-forms contain the expression $dx\,dy\,dz$.
These definitions will differ slightly in higher dimensions.

3. *Addition.* We can only add a k-form and an ℓ-form if $k = \ell$.

4. *Integration.* (a) 1-forms are integrated like line integrals.
(b) 2-forms are integrated like surface integrals. Notice that each term of the double integral in equation (2) contains a Jacobian. The variables in the Jacobian's "numerator" follows the cyclic nature of the differentials.
(c) 3-forms are integrated like ordinary triple integrals.

5. *Wedge products.* Different forms may be multiplied. In our discussion, you can only multiply a k-form and an ℓ-form if $0 \le k + \ell \le 3$. Be aware that $\omega \wedge \mu = (-1)^{k\ell}(\mu \wedge \omega)$. This property is called anticommutativity. Notice that $dx \wedge dx = dy \wedge dy = dz \wedge dz = 0$. The other properties are similar to those of real number multiplication.

6. *Differentiation.* If f is a 0-form, df is like the ordinary derivative. The "sum" rule (linearity) remains the same as in one-variable calculus. The "product" rule differs slightly and $d(d\omega) = 0$; this last identity captures the two identities $\nabla \times \nabla f = \mathbf{0}$ and $\nabla \cdot (\nabla \times \mathbf{F}) = 0$ in one formula.

SOLUTIONS TO SELECTED EXERCISES

1. (d) Since this is a 2-form wedged with a 1-form, the product is a (2+1)-form. By the distributive property, we get

$$\begin{aligned}\omega \wedge \mu &= (xy\,dy\,dz + x^2\,dx\,dy) \wedge (dx + dz)\\ &= xy\,dy\,dz \wedge dx + x^2\,dx\,dy \wedge dx + xy\,dy\,dz \wedge dz + \\ &\qquad x^2\,dx\,dy \wedge dz\\ &= xy\,(dy\,dz \wedge dx) + x^2(dx\,dy \wedge dx) + xy(dy\,dz \wedge dz) + \\ &\qquad x^2(dx\,dy \wedge dz)\\ &= xy\,dx\,dy\,dz + x^2\,dx\,dy\,dz = (xy + x^2)\,dx\,dy\,dz.\end{aligned}$$

We used the anticommutativity fact that $dy\,dz \wedge dx = (-1)^2(dx \wedge dy\,dz)$. Also, we used the associativity property along with the identities $dx \wedge dx = dz \wedge dz = 0$.

3. (b) When we differentiate this 1-form, we should get a 2-form.

$$\begin{aligned}d\omega &= [d(y^2 \cos x) \wedge dy + (-1)^0(y^2 \cos x) \wedge d(dy)] + \\ &\qquad [d(xy) \wedge dx + (-1)^0(xy) \wedge d(dx)] + d(dz)\end{aligned}$$

$= [(-y^2 \sin x\, dx + 2y \cos x\, dy) \wedge dy + 0] + [(y\, dx + x\, dy) \wedge dx + 0] + 0$
$= -y^2 \sin x\, dx\, dy - x\, dx\, dy = (-y^2 \sin x - x)\, dx\, dy$

since $dx \wedge dx = dy \wedge dy = 0$ and $dy \wedge dx = -dx\, dy$.

(e) When we differentiate this 2-form, we should get a 3-form.

$$d\omega = d[(x^2 + y^2)dy] \wedge dz + (-1)^1(x^2 + y^2)dy \wedge d(dz)$$
$$= [d(x^2 + y^2) \wedge dy + (-1)^0(x^2 + y^2) \wedge d(dy)] \wedge dz + 0$$
$$= [(2x\, dx + 2y\, dy) \wedge dy] \wedge dz = [2x\, dx \wedge dy + 2y\, dy \wedge dy] \wedge dz$$
$$= 2x\, dx\, dy\, dz$$

since $d(dz) = d(dy) = dy \wedge dy = 0$.

4. Notice that $dx \wedge dx\, dy = dy \wedge dx\, dy = dy \wedge dy\, dz = dz \wedge dy\, dz = dz \wedge dz\, dx = dx \wedge dz\, dx = 0$. The derivative of $F\, dx\, dy$ is

$$\left[\frac{\partial F}{\partial x} dx + \frac{\partial F}{\partial y} dy + \frac{\partial F}{\partial z} dz\right] \wedge dx\, dy + (-1)^0 F \wedge d(dx\, dy)$$
$$= \frac{\partial F}{\partial x} dx \wedge dx\, dy + \frac{\partial F}{\partial y} dy \wedge dx\, dy + \frac{\partial F}{\partial z} dz \wedge dx\, dy = \frac{\partial F}{\partial z} dz \wedge dx\, dy$$
$$= \frac{\partial F}{\partial z} dz \wedge (dx \wedge dy) = \frac{\partial F}{\partial z} (dz \wedge dx) \wedge dy$$
$$= -\frac{\partial F}{\partial z} (dx \wedge dz) \wedge dy = -\frac{\partial F}{\partial z} dx \wedge (dz \wedge dy)$$
$$= \frac{\partial F}{\partial z} dx \wedge (dy \wedge dz) = \frac{\partial F}{\partial z} dx\, dy\, dz.$$

Similarly, the derivative of $G\, dy\, dz$ is $(\partial G/\partial x)\, dx\, dy\, dz$ and the derivative of $H\, dz\, dx$ is $(\partial H/\partial y)\, dx\, dy\, dz$. Therefore,

$$d\eta = \left(\frac{\partial G}{\partial x} + \frac{\partial H}{\partial y} + \frac{\partial F}{\partial z}\right) dx\, dy\, dz = (\operatorname{div} \mathbf{V})\, dx\, dy\, dz.$$

8. By a direct calculation, we note that ∂S is the unit circle in the xy plane. The parametrization is $(x, y, z) = (\cos t, \sin t, 0), 0 \le t \le 2\pi$. Therefore,

$$\int_{\partial S} \omega = \int_0^{2\pi} [(\cos t + \sin t)\, 0 + (\sin t + 0)(-\sin t\, dt) + (\cos t + 0)(\cos t\, dt)]$$
$$= \int_0^{2\pi} (-\sin^2 t + \cos^2 t) dt = \int_0^{2\pi} \cos 2t\, dt = \left.\frac{\sin 2t}{2}\right|_0^{2\pi} = 0.$$

By Stokes' theorem, the above calculation is equal to $\int_S d\omega$. We compute $d\omega$:

$d\omega = [d(x + y) \wedge dz + (x + y) \wedge d(dz)] + [d(y + z) \wedge dx + (y + z) \wedge d(dx)] + [d(x + z) \wedge dy + (x + z) \wedge d(dy)] = [(dx + dy) \wedge dz] + [(dy + dz) \wedge dx] + [(dx + dz) \wedge dy]$ because $d(dx) = d(dy) = d(dz) = 0$. This simplifies to $dx \wedge dz + dy \wedge dz + dy \wedge dx + dz \wedge dx + dx \wedge dy + dz \wedge dy = -dz\, dx + dy\, dz - dx\, dy + dz\, dx + dx\, dy - dy\, dz = 0$. Therefore, we have $\int_S \omega = 0$, also.

11. By Stokes' theorem, we have $\int_{\partial R} \omega = \int_R d\omega$. We will look at the right-hand side:

$d(x\,dy\,dz + y\,dz\,dx + z\,dx\,dy)$

$= d[(x\,dy \wedge dz) + (y\,dz \wedge dx) + (z\,dx \wedge dy)]$

$= [d(x\,dy) \wedge dz + (-1)^1(x\,dy \wedge d(dz))] + [d(y\,dz) \wedge dx + (-1)^1(y\,dz \wedge d(dx))] + [d(z\,dx) \wedge dy + (-1)^1(z\,dx \wedge d(dy))]$

$= [dx \wedge dy + x \wedge d(dy)] \wedge dz + [dy \wedge dz + y \wedge d(dz)] \wedge dx + [dz \wedge dx + z \wedge d(dx)] \wedge dy$.

Since $d(dx) = d(dy) = d(dz) = 0$, we get

$(dx \wedge dy) \wedge dz + (dy \wedge dz) \wedge dx + (dz \wedge dx) \wedge dy$

$= dx\,dy\,dz + dx \wedge (dy \wedge dz) + (-dx \wedge dz) \wedge dy$

$= dx\,dy\,dz + dx\,dy\,dz - dx \wedge (dz \wedge dy)$

$= 3\,dx\,dy\,dz$.

Thus, the right-hand side is $\int_R 3\,dx\,dy\,dz$. We recognize $\int_R dx\,dy\,dz$ as the volume of R. Dividing by 3 gives us the desired result.

8.R Review Exercises for Chapter 8

SOLUTIONS TO SELECTED EXERCISES

2. By Gauss' theorem, we have $\int_{\partial\Omega} \mathbf{H} \cdot d\mathbf{S} = \int_\Omega (\nabla \cdot \mathbf{H})\,dV = \int_\Omega (\text{div}\,\mathbf{H})\,dV$, so we need to show that $\text{div}[\mathbf{F} \times (\nabla \times \mathbf{G})] = (\nabla \times \mathbf{F}) \cdot (\nabla \times \mathbf{G}) - \mathbf{F} \cdot (\nabla \times \nabla \times \mathbf{G})$. By formula 9, table 3.1, section 3.5, we have $\text{div}(\mathbf{A} \times \mathbf{B}) = \mathbf{B} \cdot \text{curl}\ \mathbf{A} - \mathbf{A} \cdot \text{curl}\,\mathbf{B} = \mathbf{B} \cdot (\nabla \times \mathbf{A}) - \mathbf{A} \cdot (\nabla \times \mathbf{B})$. Now let $\mathbf{A} = \mathbf{F}$ and $\mathbf{B} = \nabla \times \mathbf{G}$. This gives the desired result since $\mathbf{a} \cdot \mathbf{b} = \mathbf{b} \cdot \mathbf{a}$.

7. (a) To show that $\mathbf{F}$ is conservative, the easiest thing to do is to show that $\nabla \times \mathbf{F} = \mathbf{0}$:

$$\nabla \times \mathbf{F} = \begin{vmatrix} \mathbf{i} & \mathbf{j} & \mathbf{k} \\ \dfrac{\partial}{\partial x} & \dfrac{\partial}{\partial y} & \dfrac{\partial}{\partial z} \\ 6xy(\cos z) & 3x^2(\cos z) & -3x^2y(\sin z) \end{vmatrix}$$

$$= (-3x^2\sin z + 3x^2\sin z,\ 6xy\sin z - 6xy\sin z,\ 6xy\cos z - 6xy\cos z)$$

$$= \mathbf{0}.$$

(b) If $\mathbf{F}$ is the gradient of some $f(x, y, z)$, then $f(x, y, z)$ must satisfy

$6xy \cos z = \partial f/\partial x$ (1)
$3x^2 \cos z = \partial f/\partial y$ (2)
$-3x^2 \sin z = \partial f/\partial z$. (3)

Integration of (3) with respect to z gives $f(x, y, z) = 3x^2y \cos z + g(x, y)$.
Integration of (2) with respect to y gives $f(x, y, z) = 3x^2y \cos z + h(x, z)$.
Integration of (1) with respect to x gives $f(x, y, z) = 3x^2y \cos z + k(y, z)$.
Comparing the three f 's , we can see that $f(x, y, z) = 3x^2y \cos z + C$.

(c) Since **F** is a gradient, we only need to evaluate f at the endpoints to calculate the line integral. We get

$$\int_{\boldsymbol{\sigma}} \mathbf{F} \cdot d\mathbf{s} = f(\boldsymbol{\sigma}(\pi/2)) - f(\boldsymbol{\sigma}(0)) = 3(\cos^3\theta)^2(\sin^3\theta)\Big|_0^{\pi/2} = 0 .$$

9. We want to compute $\int_W \nabla \cdot \mathbf{F}\, dV$, where W is the unit cube. By Gauss' theorem,

$$\int_V \nabla \cdot \mathbf{F}\, dV = \int_0^1 \int_0^1 \int_0^1 (6z + x^2 + y)\, dx\, dy\, dz$$
$$= \int_0^1 \int_0^1 \left[\frac{1}{3} + (6z + y)\right] dy\, dz$$
$$= \int_0^1 \left(\frac{1}{3} + 6z + \frac{1}{2}\right) dz = \left(\frac{1}{3} + \frac{1}{2} + 3\right) = \frac{23}{6} .$$

13. (a) We compute $\nabla f = (\partial f/\partial x)\mathbf{i} + (\partial f/\partial y)\mathbf{j} + (\partial f/\partial z)\mathbf{k} = 3y \exp(z^2)\mathbf{i} + 3x \exp(z^2)\mathbf{j} + 6xyz \exp(z^2)\mathbf{k}$.

(b) ∇f is a gradient, so the integral is $f(\boldsymbol{\sigma}(\pi)) - f(\boldsymbol{\sigma}(0))$. Since $f(x, y, z) = 3xy \exp(z^2)$, we have $f(\boldsymbol{\sigma}(t)) = 3(3 \cos^3 t)(\sin^2 t) \exp(e^{2t})$; so $f(\boldsymbol{\sigma}(\pi)) = 0$ and $f(\boldsymbol{\sigma}(0)) = 0$. Therefore, $\int_{\boldsymbol{\sigma}} \nabla f \cdot d\mathbf{s}$ is 0 .

(c) Let A be the region whose boundary is $\boldsymbol{\sigma}(t)$. Then, by Stokes' theorem , $\int_{\boldsymbol{\sigma}} \nabla f \cdot d\mathbf{s} = \int_A [\nabla \times (\nabla f)] \cdot d\mathbf{S}$. This is always 0 because the curl of a gradient is always 0 ; the other part of Stokes' theorem has been illustrated in part (b).

14. By Green's theorem, $\int_C x^3\, dy - y^3\, dx = \int\int_D (3x^2 + 3y^2)\, dx\, dy$, where D is the unit disc. Now use polar coordinates: $x = r \cos \theta$, $y = r \sin \theta$. The unit circle is described by $0 \le r \le 1$ and $0 \le \theta \le 2\pi$. Since the Jacobian is r , the integral becomes

$$\int_0^{2\pi} \int_0^1 3r^2 \cdot r\, dr\, d\theta = 2\pi \int_0^1 3r^3\, dr = 2\pi \cdot \frac{3}{4} = \frac{3}{2}\pi .$$

17. By Green's theorem,

$$\text{Area} = \iint_D dx\,dy = \frac{1}{2}\int_C (x\,dy - y\,dx)$$

$$= \frac{1}{2}\int_0^\pi (a \sin\theta \cos\theta)(2a \sin\theta \cos\theta)\,d\theta - \frac{1}{2}\int_0^\pi (a \sin^2\theta)[a(\cos^2\theta - \sin^2\theta)]d\theta$$

$$= \frac{1}{2}a^2\int_0^\pi (2\sin^2\theta\cos^2\theta - \sin^2\theta\cos^2\theta + \sin^4\theta)d\theta$$

$$= \frac{1}{2}a^2\int_0^\pi \sin^2\theta(\cos^2\theta + \sin^2\theta)\,d\theta = \frac{a^2}{2}\int_0^\pi \sin^2\theta\,d\theta$$

$$= \frac{a^2}{2}\int_0^\pi \left(\frac{1-\cos 2\theta}{2}\right)d\theta$$

$$= \frac{a^2}{2}\left(\frac{\theta}{2} - \frac{\sin 2\theta}{4}\right)\Bigg|_0^\pi = \frac{\pi a^2}{4}.$$

20. (c) If $\mathbf{F}$ is a gradient, then $\text{curl}\,\mathbf{F} = \mathbf{0}$. We compute

$$\text{curl}\,\mathbf{F} = \begin{vmatrix} \mathbf{i} & \mathbf{j} & \mathbf{k} \\ \partial/\partial x & \partial/\partial y & \partial/\partial z \\ 2xy^3 & x^2z^3 & 3x^2yz^2 \end{vmatrix} = -6xyz^2\mathbf{j} + (2xz^3 - 6xy^2)\mathbf{k}.$$

Except at the origin, $\text{curl}\,\mathbf{F}$ is not $\mathbf{0}$, so $\mathbf{F}$ is not a gradient.

CHAPTER 9
SAMPLE EXAMS

9.1 Comprehensive Test for Chapters 1 - 4

1. (a) Find the relative extrema of $f(x, y) = x^2 - 2x + y^2 - 1$.
   (b) For the same f, find the relative extrema on the curve $x^2 + y^2 = 1$.

2. (a) Fido is on a strange mountain described by $M(x, y, z) = e^{-z} xy \sin z$. If Fido is at $(1, 1, \pi/2)$, what is the direction Fido must take to get down the mountain as fast as possible?
   (b) Fido has to follow a certain trail of tree trunks (which he has previously marked) along $\mathbf{c}(t) = (2t, t^2, 1)$. How fast is Fido's altitude changing at $t = 1$?

3. (a) Given the vector field $\mathbf{F}(x, y) = (y, -x)$, find an expression for a flow line $\boldsymbol{\sigma}(t)$ for $\mathbf{F}$ by taking a good guess.
   (b) Do the same for $\mathbf{F}(x, y) = (2x, -y)$.

4. (a) Find the arc length of $\boldsymbol{\sigma}(t) = (2t, t^2)$, $0 \le t \le 1$.
   (b) Find the arc length of $\boldsymbol{\sigma}(t) = (2\sin t, \sin^2 t)$, $0 \le t \le \pi/2$.
   (c) Explain your answers.

5. A wooden box is to be made with $120.00 worth of wood. The lid is to be made from wood that costs $2.00 per square inch, and the rest of the box is to be make from wood that costs $2.50 per square inch. What is the biggest box that could be made?

6. (a) Find an equation of the line of intersection of the planes $3x + 2y - z = 7$ and $x - 4y + 2z = 0$.
   (b) Find the equation of the plane which contains the points $(1, 3, -2)$ and $(0, -2, 1)$ and is perpendicular to the plane $3x - y - 2z = 5$.

7. Let $f: B \subset \mathbf{R}^2 \to \mathbf{R}^3$ be defined by $(u, v) \to (uv^2, v^3 - u, v \sin u)$, and $g: B \subset \mathbf{R}^2 \to \mathbf{R}^3$ be defined by $(x, y, z) \to (yze^x, y^3 \cos xz)$.
   (a) Calculate $\mathbf{D}(f \circ g)(0, 1, 0)$.
   (b) Calculate $\mathbf{D}(g \circ f)(0, 1)$.

8. An aquarium manufacturer has advertised that its tanks hold exactly $0.49\ m^3$ of water. Due to production error, the tanks have dimensions $1.01\ m \times 0.72\ m \times 0.67\ m$ rather than the specified $1.00\ m \times 0.70\ m \times 0.70\ m$. Use the linear approximation to estimate the error in the advertised capacity of the tank.

9. A budding bioengineering student who is an amateur ornithologist wanted to fly south for Christmas vacation, but couldn't afford an airline ticket. He knew that his pet swallows also wanted to fly south for the winter, so his plan was to tie his feathered friends to a hang glider and have them fly him home. His well-trained swallows had a velocity vector of $3\mathbf{i} + 4\mathbf{j} + \mathbf{k}$ (km/hr) until they reached their cruising altitude at $(0, 0, 1/2)$. At that point, they continued along the same path, but with no change in altitude. (Assume the Earth is flat.)
   (a) What was the original starting point at $(x, y, 0)$?
   (b) How long did it take to reach the cruising altitude?
   (c) When they reached $(0, 0, 1/2)$, a strong wind with velocity $2\mathbf{i} + \mathbf{j}$ affected the flight plan. What was the swallows' velocity vector to maintain a total velocity of $3\mathbf{i} + 4\mathbf{j}$?
   (d) After 3 hours, the bioengineer calculates that they will pass over his worst enemy's home, where some of his pets will give a 21-bird dropping salute. What is the position of the enemy's house?
   (e) Assuming that the swallows were unable to compensate for the wind, how far from the target point would they be after 3 hours?

10. Tightwad Terry, the miser, has performed a computer analysis of his earning power. He has determined that his earning function is $\$(t, u, v, w) = t^2 e^u + v \sin w$.
   (a) Compute $\$_t$, $\$_u$, $\$_v$, and $\$_w$.
   (b) Assuming Tightwad Terry wants to increase his earnings as fast as possible and he was at $(1, 0, 2, 0)$ on his \$ graph, how should he change t, u, v, and w ?
   (c) If he has exactly one unit of each resource (represented by t, u, v, and w), what is the largest increase in earning that he can expect? Note that he can use fractional amounts of resources, such as 1/2 of t and 1/2 of w .

9.2 Comprehensive Test for Chapters 5 - 8

1. (a) Evaluate
$$\iint_D 3(x + y)\, e^{(x - y)}\, dx\, dy$$
over the region bounded by the lines $x + y = 1$, $x + y = -1$, $x - y = 1$, and $x - y = -1$.
   (b) Evaluate
$$\int_0^1 \int_x^1 \sin y^2\, dy\, dx .$$

2. A block has a slanted top described by $x + y + z = 2$. Its edges are perpendicular to the xy plane, and the bottom of the block is formed by the triangle with vertices $(1, 0, 0)$, $(0, -1, 0)$, and $(0, 1, 0)$. What is the volume of this block?

3. Find the mass of a wall described by $0 \le y \le -x^2 - 2x + 3$, $-3 \le x \le 3$, having density $\rho(x, y) = 2|y| + 3$.

4. Let $\mathbf{F}(x, y) = (e^x \cos 3y\ , -3e^x \sin 3y)$.
   (a) Find an $f(x, y)$ such that $\nabla f = \mathbf{F}$ for all (x, y) .
   (b) Evaluate $\int_{\sigma} \mathbf{F} \cdot d\mathbf{s}$ for the path $\boldsymbol{\sigma}(t) = (\cos t\ , \sin t)\ , 0 \le t \le 2\pi$.
   (c) Compute $\nabla \times \mathbf{F}$.
   (You should be able to do parts (b) and (c) in your head.)

5. (a) A certain sugar solution is slowly permeating though a membranous surface described by $x^2 + y^2 + 2z^2 = 1$ at the velocity $2x\mathbf{i} + 3y\mathbf{j} + z\mathbf{k}$. How much of the sugar solution flows through the membrane in a unit time?
   (b) A particle moves in a path $\boldsymbol{\sigma}(t) = (2t, 3t, t)$ in a force field $\mathbf{F} = (2x, 2y, 4z)$. What is the work done in the time interval $0 \le t \le 5$?

6. (a) A contact lens can be described as a cap of a sphere of radius R cut out by a cone of angle $\pi/4$. Find the surface area of the lens. (Hint: Set up the lens in a correct and convenient coordinate system)
   (b) 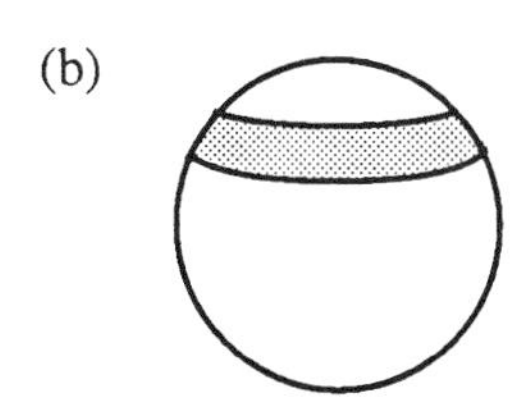
   Slice a sphere anywhere with 2 parallel planes which are separated by a fixed distance d . The bands obtained always have the same surface area. Prove this. What is the area? (Hint: Use spherical coordinates. What is Δz ?)

7. (a) Find the volume enclosed by the surface
   $$x^2 + 9y^2 + z^2 - 2x + 54y - 10z + 106 = 0\ .$$
   (b) Find an expression for the surface area of the surface above.

8. Which of the following vectors are conservative?
   (a) $\mathbf{F}(x, y, z) = (x^2 + 1/x\ , \ln (y + 1)\ , z)$
   (b) $\mathbf{F}(x, y, z) = \left(\dfrac{-3x}{(x^2 + y^2 + z^2)^{3/2}}, \dfrac{-3y}{(x^2 + y^2 + z^2)^{3/2}}, \dfrac{-3z}{(x^2 + y^2 + z^2)^{3/2}} \right)$
   (c) $\mathbf{F}(x, y, z) = (3yz, 2xz, 5xy)$

9. Complete the following statement: A vector field $\mathbf{F}$ is conservative if _____ .
   (i) there is a vector field $\mathbf{G}$ such that $\mathbf{F} = \text{curl } \mathbf{G}$.
   (ii) there is a scalar function f such that $\nabla f = \mathbf{F}$.
   (iii) $\text{div } \mathbf{F} = 0$.

   (a) (i) only (b) (i) and (ii) (c) (ii) only (d) (ii) and (iii) .

10. Consider the following argument: given a vector field $\mathbf{F}$,

$$\int_S (\nabla \times \mathbf{F}) \cdot d\mathbf{S} = \int_{\partial S} \mathbf{F} \cdot d\mathbf{s} \qquad \text{(by Stokes' theorem)}$$

but

$$\int_V \nabla \cdot (\nabla \times \mathbf{F})\, dV = \int_{S=\partial V} (\nabla \times \mathbf{F}) \cdot d\mathbf{S} \qquad \text{(by Gauss' theorem)}$$

so everything is 0 because the divergence of the curl of any vector field is 0. What's wrong?

9.3 Comprehensive Test A for Chapters 1 - 8

1. If true, state true. If false, explain why.

   (a) The path integral $\int_\sigma 2\pi\, ds$ is the surface area of a cylinder of radius 1 and height 2π if $\sigma = (\cos t, \sin t, t)$, $0 \le t \le 2\pi$.
   (b) A smooth function $f(x, y)$ with at least two critical points must have both a relative minimum and a relative maximum.
   (c) The line integral of a mass density function along a curve is the mass of the curve.
   (d) curl $\mathbf{F} = \nabla \cdot \mathbf{F}$.
   (e) $\partial^2 f/\partial x \partial z = \partial^2 f/\partial z \partial x$ for all continuous functions $f(x, y, z)$.

2. Multiple choice. Choose the one correct answer.

   (1) Which pair of vectors have the smallest angle between them?
   (a) $\mathbf{i} - \mathbf{j}, \mathbf{j} + 2\mathbf{k}$
   (b) $3\mathbf{i} + 2\mathbf{k}, -2\mathbf{j}$
   (c) $2\mathbf{i} - \mathbf{j} + \mathbf{k}, \mathbf{i} - \mathbf{j} + \mathbf{k}$
   (d) There is insufficient information.

   (2) Green's theorem requires
   (i) continuous first partial derivatives
   (ii) continuous functions
   (iii) any closed curve which is the boundary of a region

   (a) i only (b) ii only (c) ii and iii (d) i and iii.

   (3) Let $f(x, y, z) = z^2 x\, e^x \cos(yz)$. A particle travels along a path σ from $(0, -2/5, \pi/4)$ to $(3, 5, -\pi/2)$. If the force acting on the particle is ∇f, then the work done on the particle is
   (a) negative
   (b) zero
   (c) positive
   (d) unknown since σ is unknown

(4) For any vectors **u** and **v**, $\mathbf{u} \cdot (\mathbf{u} \times \mathbf{v}) =$

(i) $\mathbf{v} \cdot (\mathbf{u} \times \mathbf{v})$
(ii) 0
(iii) $(\mathbf{v} \times \mathbf{u}) \cdot \mathbf{u}$

(a) 1 only (b) 2 only (c) 1 and 2 (d) 1, 2 and 3 .

3. Let $\mathbf{F} = x\mathbf{i} + y\mathbf{k}$. Compute $\int_S \text{curl } \mathbf{F} \cdot d\mathbf{S}$ for the following surfaces S :
(a) $x^2 + y^2 + (z-3)^2 = 1,\ z \geq 3$
(b) $9x^2 + 9y^2 + z^2 = 9,\ z \geq 0$

4. Consider the system
$$F_1(u, v, x, y) = u^2 - v^2 + 2x + xy$$
$$F_2(u, v, x, y) = 2u + 3v - 5x^2$$
(a) Show that one can not solve for u and v in terms of x and y at $(u, v) = (0, 0)$.
(b) Show that $\partial u/\partial x$ exists at $(u, v, x, y) = (3, 2, -1, 0)$. Compute it.

5. Suppose $g(u, v) = (uv^2, u + 2v, v)$. At $(x, y, z) = (0, 0, 0)$, $(u, v) = (1, 2)$, and $\mathbf{D}f = \begin{bmatrix} 1 & -1 & 0 \\ 2 & 3 & 1 \end{bmatrix}$. Let $h = g \circ f$. What is $\mathbf{D}h(0, 0, 0)$?

6. A surface is parametrized by
$$x = u^2,\quad y = v^2,\quad z = uv,\quad 0 \leq u \leq 2,\quad 0 \leq v \leq 2 .$$
(a) Find the tangent plane at $(u, v) = (1, 1)$ as a function of x and y .
(b) When $u = 1$, find the arc length of the curve in space for $0 \leq v \leq 2$.

7. Let $\mathbf{\Phi}$ be the surface parametrized by
$$x = e^u \cos v,\ y = e^u \sin v,\ z = v,\ 0 \leq u \leq 1,\ 0 \leq v \leq \pi/2 .$$
(a) Find the area of the surface $\mathbf{\Phi}$.
(b) Compute the average of z over $\mathbf{\Phi}$.

8. Let $f(x, y) = x^2 - 3x + y - y^2 + 5$.
(a) At all critical points (x_0, y_0), express $(x_0, y_0, f(x_0, y_0))$ in spherical coordinates.
(b) Find the extrema of f on the circle $x^2 + y^2 = 5/2$.

9. A surface is described by the equation $z = 2y + \cos \pi x - \sqrt{x}$. Consider the point $(4, -1, 1)$.
(a) Find the directional derivative in the direction of $\mathbf{i} + 2\mathbf{j}$.
(b) Fine the equation of the tangent plane.

10. A region in space lies between the graphs of $z = 16 + x^2 + y^2$ and $z = \sqrt{x^2 + y^2}$. The region also lies inside the cylinder $x^2 + y^2 = 4$.
    (a) Express the volume of the region as a triple integral with cartesian coordinates.
    (b) Rewrite your answer in (a) with cylindrical coordinates.
    (c) Find the volume of the region.

9.4 Comprehensive Test B for Chapters 1 - 8

1. Physical and geometric interpretations.
    (a) What do you know about $\mathbf{u}$ and $\mathbf{v}$ if $\mathbf{u} \cdot \mathbf{v} = 0$?
    (b) What do you know about $\mathbf{u}$ and $\mathbf{v}$ if $\mathbf{u} \times \mathbf{v} = \mathbf{0}$?
    (c) What physical interpretation is associated with a negative divergence?
    (d) Give a physical interpretation of curl $\mathbf{V}$.

2. Consider the point $(1, 1, 1, 1)$ of the function $f(x, y, z, t) = x^2y + zt - z$.
    (a) Compute the directional derivative in the direction of $(2, 0, 6, 4)$.
    (b) Is f increasing faster in the direction of $(2, 0, 6, 4)$ or in the direction of $(1, 1, 2, 1)$? Explain your answer.

3. (a) The cylinder $x^2 + y^2 = 4$ is cut by the plane $x + y + z = 1$. Show that the arc length of the intersecting curve is

$$\sqrt{8}\int_0^{2\pi} \sqrt{1 - \cos\theta \sin\theta}\, d\theta .$$

    (b) For the intersecting curve, find an equation for the tangent line at $(1, \sqrt{3}, -\sqrt{3})$.

4. Find the minimum and maximum of $x + yz$ subject to the constraints $x + z = 2y$ and $\{(x, y) \mid (x, y) \in [3, 5] \times [0, 2]\}$.

5. Let S be the boundary of a box $B = [-2, 2] \times [-1, 1] \times [-3, 3]$, $\mathbf{F} = 2x\mathbf{i} + 3z\mathbf{j} + 2y\mathbf{k}$, and $\mathbf{G} = x^3\mathbf{i} + 3z\mathbf{j} + 2y\mathbf{k}$.
    (a) Compute the integral of $\nabla \cdot \mathbf{F}$ over B.
    (b) Compute $\int_S \mathbf{G} \cdot d\mathbf{S}$.
    (c) Suppose the origin at the center of B is shifted to $(8, -15, 20)$ and then rotated $30°$ around the y axis. Compute $\int_S \mathbf{F} \cdot d\mathbf{S}$.

6. A hole of radius $1/2$ is drilled through the axis of symmetry of the hemisphere $x^2 + y^2 + z^2 = 1$, $z \geq 0$.
    (a) Write the volume of the remaining piece in cartesian coordinates.
    (b) Write the volume of the remaining piece in cylindrical coordinates.
    (c) Compute the volume.

7. 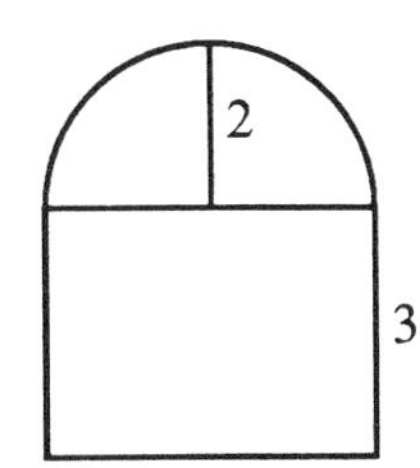

A painter is scared of heights, so he charges z^2 dollars per square meter to paint objects located at height z. His latest job is to paint the silo S shown here. The height of the cylindrical part is 3 meter and the radius of the hemispherical part is 2 meter.

(a) Compute $\int_S dS$ and interpret geometrically.

(b) Compute how much the painter charges.

8. (a) Compute $\int_0^1 \int_x^1 \cos(y^2 + 3)\, dy\, dx$.

(b) Compute $\int_0^\infty \int_0^\infty \int_0^\infty e^{-(x^2+y^2+z^2)^{1/2}}\, dx\, dy\, dz$.

9. (a) Verify Stokes' theorem for $\mathbf{F} = z^3\mathbf{i} + (x^3 - y^3)\mathbf{j} + y^3\mathbf{k}$ over the hemisphere $x^2 + y^2 + z^2$, $z \geq 0$.

(b) For the same $\mathbf{F}$ as in (a), evaluate $\int_S (\nabla \times \mathbf{F}) \cdot d\mathbf{S}$ for the surface $x^2 + y^2 + 5z^2$, $z \leq 0$.

10. (a) Find a vector-valued function $f(x, y, z)$ such that

$$\mathbf{D}f(x, y, z) = \begin{bmatrix} -yz \sin(xy)\, e^{\cos(xy)} & -xz \sin(xy)\, e^{\cos(xy)} & e^{\cos(xy)} \\ y^2 \sin z & 2xy \sin z & xy^2 \cos z \end{bmatrix}.$$

(b) 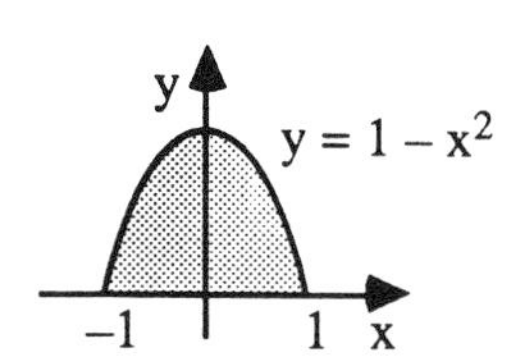

For the region D shown at the left, let V be the volume of the solid lying between $f(x, y) = x^3 \sin y$ and the xy plane and lying over D. Write V in the form $\iiint g(x, y, z)\, dz\, dy\, dx$.

(c) Rewrite your answer to part (b) in the form $\iiint g(x, y, z)\, dz\, dx\, dy$.

APPENDIX
ANSWERS TO SAMPLE EXAMS

A.1 Answers to Comprehensive Test for Chapters 1 - 4

1. (a) (1, 0) is a minimum
   (b) (1, 0), (–1, 0)

2. (a) $e^{-\pi/2}(-1, -1, 1)$
   (b) $6e^{-1} \sin 1$

3. (a) $\boldsymbol{\sigma}(t) = (\cos t, \sin t)$

   (b) $\boldsymbol{\sigma}(t) = (\exp(2t), \exp(-t))$

4. (a) $\sqrt{2} + \ln(1 + \sqrt{2})$
   (b) $\sqrt{2} + \ln(1 + \sqrt{2})$
   (c) The two parametrizations actually trace out the same path.

5. $32\sqrt{5/3}$ cubic inches

6. (a) (2, 1, 1) + t(0, 1, 2)
   (b) 13x + 7y + 16z = 2

7. (a) $\begin{bmatrix} 0 & 0 & 1 \\ 0 & 9 & -1 \\ 0 & 0 & 1 \end{bmatrix}$ (b) $\begin{bmatrix} 1 & 0 \\ -3 & 9 \end{bmatrix}$

8. $0.0021\ m^3$

9. (a) (–3/2, –2) (b) 1/2 hour
   (c) $\mathbf{i} + 3\mathbf{j}$ (d) (15/2, 10, 0)
   (e) $5\sqrt{5}/2$ km

10. (a) $\$_t = 2te^u$, $\$_u = t^2 e^u$, $\$_v = \sin w$, $\$_w = v \cos w$
    (b) Go in the direction of (2, 1, 0, 2)
    (c) $(2e^{1/3}/3, 4e^{1/3}/9, 0, 0)$

A.2 Answers to Comprehensive Test for Chapters 5 - 8

1. (a) 0 (b) (1 – cos 1)/2

2. 5/3

3. 1848/5

4. (a) $f(x, y) = e^x \cos 3y + C$
   (b) 0 (c) **0**

5. (a) $-16\pi/3\sqrt{2}$ (b) 375

6. (a) $2\pi R^2 (1 - \sqrt{2}/2)$
   (b) The area is $2\pi R^2 d$

7. (a) $4\pi/9$
   (b) $\frac{1}{3} \int_0^{2\pi} \int_0^{\pi} \sin \varphi \sqrt{1 + 8\sin^2\varphi \sin^2\theta} \, d\varphi \, d\theta$

8. (a) and (c)

9. (c)

10. In Stokes' theorem, the surface should not be closed (so it would have a real boundary), whereas in Gauss' theorem, the surface has to be closed (to enclose a volume).

A.3 Answers to Comprehensive Test A for Chapters 1 - 8

1. (a) False. The surface area of such a cylinder is 2π, while the value of the integral is $2\sqrt{2}\,\pi$.
   (b) False. They could both be saddle points, for example.
   (c) True
   (d) False. The curl is a cross product, not a dot product.
   (e) False. We also need continuously differentiable first partials.

2. (1) (c) (2) (d) (3) (d) (4) (d)

3. (a) 0 (b) 0

4. (a) $\begin{vmatrix} \frac{\partial F_1}{\partial u} & \frac{\partial F_1}{\partial v} \\ \frac{\partial F_2}{\partial u} & \frac{\partial F_2}{\partial v} \end{vmatrix} = \begin{vmatrix} 0 & 0 \\ 2 & 3 \end{vmatrix} = 0$

   (b) $\partial u/\partial x = -2$

5. $\begin{bmatrix} 12 & 8 & 4 \\ 5 & 5 & 2 \\ 2 & 3 & 1 \end{bmatrix}$

6. (a) $x + y - 2z = 0$
   (b) $(1/2)[4\sqrt{17} + \log(4 + \sqrt{17})]$

7. (a) $(\sqrt{2}/2)\pi(e - 1)$ (b) $1/2$

8. (a) $(\sqrt{46}/2, \tan^{-1}(1/3), \cos^{-1}(6/\sqrt{46}))$
   (b) $(3/2, 1/2), (-9/10, 13/10)$

9. (a) $15/4$ (b) $x - 8y + 4z = 16$

10. (a) $4\int_0^2\int_0^{\sqrt{2-x^2}} 16 + x^2 + y^2 - \sqrt{x^2 + y^2}\, dy\, dx$

    (b) $2\pi\int_0^2\int_0^r r\, dz\, dr + 2\pi\int_2^{16}\int_0^2 r\, dr\, dz + 2\pi\int_0^4\int_0^{r^2} r\, dz\, dr$

    (c) $496\pi/3$

A.4 Answers to Comprehensive Test B for Chapters 1 - 8

1. (a) The two vectors are orthogonal, or one of them is $\mathbf{0}$.
   (b) The two vectors are parallel, or one of them is $\mathbf{0}$.
   (c) There's more material going in than out; i.e., we have a sink.
   (d) curl $\mathbf{V} \cdot \mathbf{n}$ tells you about the circulation of $\mathbf{V}$ on the surface with normal $\mathbf{n}$.

2. (a) 8
   (b) In the direction of $(1, 1, 2, 1)$ because the directional derivative using normalized vectors is larger in the second case.

3. (a) The intersecting curve can be parametrized by
   $$(x, y, z) = (2\cos\theta, 2\sin\theta, 1 - 2\cos\theta - 2\sin\theta).$$
   (b) $(x, y, z) = (1, \sqrt{3}, -\sqrt{3}) + t(-\sqrt{3}, 1, \sqrt{3} - 1)$

4. Maximum = 6, minimum = 3

5. (a) 96 (b) 192
   (c) 96

6. (a) $4\int_{1/2}^{1}\int_{\sqrt{\frac{1}{4}-x^2}}^{\sqrt{1-x^2}}\int_{0}^{\sqrt{1-x^2-y^2}} dz\, dy\, dx$

   (b) $\int_{0}^{2\pi}\int_{0}^{1/2}\int_{0}^{\sqrt{1-r^2}} r\, dz\, dr\, d\theta$

   (c) $\frac{\pi}{3}\left(1-\left(\frac{3}{4}\right)^{3/2}\right)$

7. (a) 16π, which is the surface area of the silo.
   (b) $(500\pi/3)$ dollars

8. (a) $[\sin(4) - \sin(3)]/2$
   (b) π

9. (a) $3\pi/4$ (b) $-3\pi/4$

10. (a) $f(x, y, z) = (z\, e^{\cos(xy)}, xy^2 \sin z)$

    (b) $\int_{-1}^{1}\int_{0}^{-x^2+1}\int_{0}^{x^3 \sin y} dz\, dy\, dx$

    (c) $\int_{0}^{1}\int_{-\sqrt{1-y}}^{\sqrt{1-y}}\int_{0}^{x^3 \sin y} dz\, dx\, dy$